# LEADERS, GROUPS, AND COALITIONS

*Understanding the People and Processes in Foreign Policymaking*

This special issue of the *International Studies Review* is part of the International Studies Association's Millennial Series; each issue in this series elaborates on the theme of an annual meeting during the transition to the new century. This issue reflects the theme of the 1999 annual meeting held in Washington, D.C., entitled "One Field, Many Perspectives: Building the Foundations for Dialogue."

Editor

Margaret G. Hermann

# LEADERS, GROUPS, AND COALITIONS

## *Understanding the People and Processes in Foreign Policymaking*

| | | |
|---|---|---|
| **Preface** | | 1 |
| **Joe D. Hagan** | Does Decision Making Matter? Systemic Assumptions vs. Historical Reality in International Relations Theory | 5 |
| **Margaret G. Hermann** | How Decision Units Shape Foreign Policy: A Theoretical Framework | 47 |
| **Margaret G. Hermann** **Thomas Preston** **Baghat Korany** **Timothy M. Shaw** | Who Leads Matters: The Effects of Powerful Individuals | 83 |
| **Charles F. Hermann** **Janice Gross Stein** **Bengt Sundelius** **Stephen G. Walker** | Resolve, Accept, or Avoid: Effects of Group Conflict on Foreign Policy Decisions | 133 |
| **Joe D. Hagan** **Philip P. Everts** **Haruhiro Fukui** **John D. Stempel** | Foreign Policy by Coalition: Deadlock, Compromise, and Anarchy | 169 |
| **Ryan K. Beasley** **Juliet Kaarbo** **Charles F. Hermann** **Margaret G. Hermann** | People and Processes in Foreign Policymaking: Insights from Comparative Case Studies | 217 |

# Preface

This issue of the *International Studies Review* is part of the International Studies Association's (ISA) Millennial Series intended to assess the state of the field as we move into a new century. Each issue in this series also elaborates on the theme of an ISA annual meeting during the transition from the twentieth to the twenty-first century. The articles that follow present an overview of the study of foreign policy decision making as well as reflect the theme of the 1999 ISA convention held in Washington, D.C., that focused on "One Field, Many Perspectives: Building the Foundations for Dialogue."

Close to fifty years ago Richard Snyder and his colleagues (Snyder, Bruck, and Sapin, 1954; Snyder, 1958) introduced international relations scholars to decision making as an approach to the study of political phenomena and, more particularly, the analysis of international politics. The original monograph reviewed the literature on decision making then extant across the social sciences and showed its relevance for exploring how foreign policy decisions are made. Snyder and his associates were interested in the effects that the definition of the situation can have on who gets involved in the policymaking process and the influence of how groups and organizations are configured on the resulting foreign policy. They believed that policymakers' interpretations of the world and the ways their preferences were aggregated in the decision-making enterprise could shape what governments and institutions did. Their monograph laid the foundation for a large, and still growing, body of research that has examined how leaders, groups, and coalitions of actors can affect the way foreign policy problems are framed, the options that are selected, the choices that are made, and what gets implemented. Since its introduction, decision making has remained one perspective for understanding what is happening in international affairs.

Interest, however, in foreign policy decision making as a legitimate area of inquiry within the mainstream international relations community has ebbed and flowed across the past fifty years. For the past several decades, many scholars of world politics have discounted the need to examine the foreign policymaking process and those involved in making decisions, proposing instead to focus on the international constraints that limit what policymakers can do. The rationale went as follows. Because the systemic imperatives of anarchy or interdependence are so clear, the leadership of most governments is limited in the range of foreign policy strategies that are available to them. If these policymakers are to exercise rational leadership and maximize their country's movement toward its goals, only certain actions are feasible. Consequently, incorporating the foreign policymaking process into general theories of international relations was unnecessary since such knowledge would add little to our understanding of the dynamics of conflict, cooperation, and change in international affairs.

In the bipolar international system that characterized the Cold War, such a rationale might have seemed reasonable. But today there is little consensus on the nature of the "new world order" and more room for interpretation, innovation, misunderstanding, and miscommunication. In such an ambiguous environment, the perspectives of those involved in the foreign policymaking process can influence what governments do. Moreover, as international constraints on foreign policy become more flexible and indeterminate, the importance of domestic political concerns has increased. Scholars of international relations have begun to talk not only about different kinds of states—democracies, transitional democracies, autocracies—but also about how domestic political pressures can help to define the state—strong, weak; stable, unstable; cohesive, fragmented; satisfied, revisionist. And they have begun to emphasize that government leaders have some choice in the roles their countries play in international politics—doves, hawks; involved, isolationist; unilateral, multilateral; regional, global; pragmatic, radical. These different interpretations influence what governments do in the international arena.

With this new-found curiosity about the leadership setting, it seemed timely to examine what scholars who have spent most of their professional careers studying foreign policymaking have discovered about the conditions under which people and processes become important in shaping states' policies and activities. This special issue presents one framework that attempts to integrate and synthesize this literature. The articles explore how leaders in various types of configurations are likely to perceive and interpret the constraints in their environments, balancing international imperatives with those arising from, or embedded in, domestic politics. The lesson repeated often in this research is that international constraints only have policy implications when they are perceived as such by the leaders whose positions count in dealing with a particular problem. Whether and how these leaders will judge themselves constrained appears to depend on the nature of the domestic challenges to their leadership, how they are organized, and what they are like as people. Indeed, it would seem that to understand how the leaders in a government are likely to respond to a problem, we need to be able to demarcate which leaders and leadership groups will become more caught up in the flow of events—and, thus, perceive external forces as limiting their parameters for action—and which will instead challenge the international constraints in their path.

The articles that follow not only overview the foreign policymaking literature but they represent the report of a project that involved dialogue and discussion among scholars with particular case and regional expertise and those interested in studying how foreign policy decisions are made in a comparative context. The decision units framework that is presented here engages literature from history, social and organizational psychology, decision theory, bureaucratic politics, and comparative politics as well as that on foreign policymaking in international relations. The authors convened several times to discuss the framework and to evaluate its application to a number of historical cases. The various decision units models described in the following pages evolved from this interaction in an iterative fashion. As a result, the framework has benefited from the insights that resulted from these meetings and the ensuing dialogue and debate.

Graduate students in a series of courses on the comparative study of foreign policy at Ohio State University and the Maxwell School, Syracuse University have also analyzed cases using the framework as part of an evolving essay examining how various theoretical perspectives at different levels of analysis assist us in understanding governments' foreign policy activity. The students provided a critique of the particular decision units model they used, indicating how closely the outcome proposed by the model matched its historical counterpart and comparing the behavior indicated by the model with that resulting from other theories examined in the course. Moreover, they offered suggestions for improving the framework. A number of the students took pieces of the framework to study as dissertation topics.

Thus, the decision units framework, and this special issue, were collaborative enterprises and incorporate in a variety of ways the expertise of those who tried applying the various models to case material. The articles in this special issue are intended to (1) indicate the relevance of decision units to understanding what governments do in world affairs by examining the historical record and, more particularly, several important twentieth-century foreign policy decisions—those that led to the outbreaks of World War I, World War II, and the Cold War; (2) present the reader with an overview of the framework; (3) synthesize the research that has focused on each type of decision unit—leader, group, and coalition—and elaborate a model for each unit using cases to illustrate important variables and their linkages; and (4) evaluate the framework based on the results from sixty-five case studies using the various models. The articles are intended to be cumulative, building on one another. Thus, we encourage the reader to start at the beginning and proceed in order through the pieces; at the least, the reader should understand the nature of the framework before tackling an article on a particular type of decision unit or the evaluation of the framework.

This project received funding from the National Science Foundation (SBR-9113599); the Mershon Center at Ohio State University; the Global Affairs Institute at the Maxwell School, Syracuse University; the Milward L. Simpson

Fund at the University of Wyoming; and the Eberly College of Arts and Sciences, West Virginia University. We remain indebted to the students who completed a majority of the case studies examined here as well as the many colleagues who attended our ISA panels across the years providing us with comments and insights. We would like to dedicate this special issue in memoriam to Richard Snyder who believed a framework like that proposed here was possible, encouraged its development with both monetary and intellectual support, and inspired several generations of scholars to explore how foreign policy decisions are made.

## REFERENCES

SNYDER, RICHARD C. (1958) "Decision-Making as an Approach to the Analysis of Political Phenomena." In *Approaches to the Study of Politics*, edited by Roland Young. Evanston, IL: Northwestern University Press.

SNYDER, RICHARD C., H. W. BRUCK, AND BURTON SAPIN (1954) *Decision-Making as an Approach to the Study of International Politics*. Princeton, NJ: Princeton University Foreign Policy Analysis Project Monograph 3.

# Does Decision Making Matter?

## Systemic Assumptions vs. Historical Reality in International Relations Theory

### Joe D. Hagan
#### West Virginia University

Nearly fifty years ago Richard Snyder and his colleagues (1954) articulated a foreign policy decision-making perspective that suggested people matter in international affairs. This now classic work proposed such enduring concepts as "definition of the situation" and "organizational context." Soon recognized as a distinct "level of analysis" in the study of international relations (e.g., Singer, 1961), by the 1970s decision making had become a dominant approach in the study of foreign policymaking with the proliferation of such theoretical models as bureaucratic politics (Allison, 1971; Halperin, 1974) and groupthink (Janis, 1972) as well as a number of so-called cognitive approaches to governmental decision making (Holsti, 1976; Jervis, 1976). This prominence, however, did not last long. In the 1980s systemic theories focusing on international structures regained their former primacy with rigorously formulated neorealist (Waltz, 1979; Gilpin, 1981) and neoliberal arguments (Keohane, 1984). A common feature in this "structural realism" was the dismissal of the significance of decision-making influences and other state-level phenomena on policymaking. Waltz's critique was damning in two respects: (1) by pointing out that state-level explanations rest on reductionist logic, he portrayed foreign policy analysis as divorced from the context of international politics; and (2) by asserting that the imperatives of systemic structures (e.g., anarchy) were clear, he indicated that all leaders would readily understand them and respond according to their state's position in the system. In short, not only did decision-making approaches miss much of the substance of international politics, they also dwelled on the largely irrelevant "noise" of internal processes in explaining state behavior. Yet by the 1990s systemic realism was itself being subjected to provocative critiques, the most compelling of which involved case studies by "soft" realists that pointed out empirical gaps—puzzles and anomalies—in systemic explanations of great power foreign policies dating back to the Crimean War (e.g., Snyder, 1991; Rosecrance and Stein, 1993; Kupchan, 1994).

The premise of this article is that decision-making approaches are well positioned to contribute to the further evolution of international relations theory. This statement should not be considered a rejection of the theoretical developments of the past two decades; nor is it an assertion that decision making constitutes a comprehensive theoretical perspective or level of analysis. Rather, the argument here is that decision-making approaches can fill some of the gaps and account for some of the resulting anomalies in systemic explanations of conflict and war. The argument is made in three parts. First, the evidence concerning the origins of the twentieth century's three great power conflicts—World War I, World War II, and the Cold War—is summarized with a focus on what cannot be explained by the systemic logic of structural realism. Second, the proposal is asserted and discussed that systemic explanations are incomplete because foreign policy problems are inherently complex (see Steinbruner, 1974)—even in crisis situations where the threat of war appears imminent. To demonstrate this point, the presentation draws upon standard historical analyses of the origins of WWI, WWII, and the Cold War. Such research consistently shows that the conditions inherent in a "unitary, rational actor" model do not hold up very well. Instead, to one degree or another, leaders in the twentieth century's major crises (1) faced very real uncertainty in responding to international threats, (2) confronted trade-offs across competing goals, including that of retaining power, and (3) operated in decision structures in which political authority was quite dispersed and fragmented. This analysis indicates that understanding decision-making influences is essential to explaining how leaders will respond to international (and domestic) imperatives. Third, the article concludes by exploring the implications of the historical record for the further development of the decision-making perspective by arguing that decision-making structures, or "units," channel and focus other influences on governments and are themselves variable across international systems and domestic political structures.

## EMPIRICAL LIMITS OF SYSTEMIC EXPLANATIONS OF TWENTIETH-CENTURY GREAT POWER CONFLICTS

This section draws upon the critiques of structural-realist theory that have isolated a key limitation in systemic explanations of war—the inability to account for the outbreak of the First and Second World Wars as well as the escalation of the Cold War. Works by Snyder (1991), Rosecrance and Stein (1993), and

Kupchan (1994) as well as Levy (1990–91) have collectively demonstrated that the great powers acted in ways that cannot be readily explained as a direct response to systemic imperatives. Although not rejecting the structural-realist view of states coping with systemic anarchy, this scholarship has isolated "puzzles" in which great powers pursued policies that were inconsistent with the severity of systemic threats and/or their capabilities relative to other powers. States either *underreacted* by failing to balance against, and thereby deter, adversaries; or they *overreacted* to threats by overextending their power and/or provoking self-encirclement by other powers.

The outbreak of World War II seems to be the most divorced from the systemic premises of structural-realist theories. If ever there was a clear threat to international stability, it was Japan in East Asia and Germany in Europe by the mid-1930s. Yet the major status quo powers of the time—Britain, France, and the United States—did very little to counter the rising power of Germany and Japan.[1] These major powers failed to use their originally superior military power to reverse Japan's initial aggression in Manchuria and left unchallenged Germany's rearmament, the remilitarization of the Rhineland, and the incorporation of Austria and the Sudetenland. Nor did Britain, France, or the U.S. "balance" in the larger sense, engaging in at most slow and partial rearmament and failing to form alliances with each other and/or the historically crucial (Soviet) Russia. Ultimately, at the height of prewar tensions (the Munich crisis), Hitler was met with forceful appeasement—not deterrence—by an assertive British government, which was meekly followed by an ambivalent France and an absent United States that hitherto responded with stricter neutrality acts.[2] Also puzzling are the expansionist foreign policies of Germany and Japan. Unless one assumes the darkest image of international systems, the "revisionist" foreign policies of Germany and Japan in the 1930s are far more than a response to the security dilemmas of international anarchy. Whatever the case on this point, it is not apparent why these powers overextended themselves in

---

[1] This sketch of WWII puzzles closely follows those identified in the Snyder, Rosecrance and Stein, and Kupchan volumes. In particular, Snyder (1991: ch. 4) questions the Japanese attack on Manchuria, while Kupchan (1994: ch. 5) puzzles over the expansion of the war into China and then the Pacific. On war in Europe, Kupchan (1994: chs. 3, 4, and 7) is the most comprehensive, dealing with British, French, and U.S. appeasement before WWII. Rosecrance and Steiner (1993) also address British appeasement, while in the same volume Stein (1993) examines the underextension of American power. On the distinction between revisionist and status quo states see Schweller (1994).

[2] Even after the start of the Second World War in Europe, the French, British, and American governments persisted in, at best, weak balancing. Only with the fall of France did Britain finally abandon thoughts of an accommodation with Germany (see Lukacs, 1999) and the Roosevelt administration reverse the neutrality acts in major ways through lend-lease aid (see Holsti, 1993).

self-defeating ways. For example, even though it made sense for Japan to conquer resource-rich Manchuria, expanding the war into China in the mid-1930s and then launching a Pacific war against the United States was far beyond that government's economic and military capabilities relative to the other powers.

The occurrence of war does not depend upon actors with the extreme postures of the 1930s. The outbreak of the First World War resulted from a decade-long deterioration of the European balance-of-power system and was centered around comparatively subtle "security dilemmas" among status quo (i.e., non-revolutionary) powers. Beginning with Wilhelmine Germany's abandonment of Bismarckian restraint in favor of a militant diplomacy (itself a self-defeating behavior), by 1912 most of the increasingly vulnerable European governments had shifted to relatively hard-line foreign policies. Thus, by 1914, it is not surprising that the European states would consider war—indeed, it is arguable that the crisis was an exercise in diplomatic brinkmanship in which Austria-Hungary would quickly suppress Serbia, and Germany would threaten war to neutralize Russia and destroy the Triple Entente. Following Levy's (1990–91) analysis, what is so striking is that the attempt failed, and instead provoked a European-wide war that neither aggressor had expected, desired, or ultimately survived.[3] The failure of crisis management (or micro-level balancing) in July 1914 presents several puzzles. First, why did it take the Austro-Hungarian government nearly a month to strike at Serbia, a delay that saw European opinion shift from sympathy regarding the assassination to alarm about European stability leading to Russia's involvement in the crisis? Second, why did neither normally cautious France nor the now frustrated Germany restrain their respective allies before the crisis escalated, as they had done in previous crises over the Balkans? Third, once the threat of a wider war became apparent, why were the diplomatic initiatives of the British foreign secretary (Grey) and the German chancellor (Bethmann-Hollweg) so completely ineffective in containing the crisis, in marked contrast to the concert diplomacy of the previous Balkan

---

[3]This treatment of the July 1914 crisis as a systemic puzzle draws from Levy (1990–91) as well as Hermann and Hermann (1969). It is consistent with but not directly based on Snyder, Rosecrance and Stein, and Kupchan; none of these scholars examined the outbreak of the war but, instead, explored balancing behavior in the larger time frame. Some of the flaws in British decision making in the July 1914 crisis noted here are downplayed by Kupchan, who argues that the British and French balanced effectively before WWI. In the analysis here, the July 1914 crisis is viewed as the culmination of a "spiral of conflict" (Jervis, 1976) and represents a failure of both brinkmanship and deterrence—either of which points to a puzzling collapse of the balance of power. As such, and like much of the literature on the July crisis, this analysis falls between the image of the crisis as accidental and entirely unintended and the view that it was planned by an expansionist Germany.

crises? Finally, why did the British government fail to express in a timely manner its commitment to intervene on the side of its fellow Entente powers if Germany attacked France? The irreality of Britain's failure to deter German brinkmanship is as striking as the ineptitude of Austria-Hungary's projection of its military force against Serbia.

In many respects the Cold War is not puzzling. It can be argued that the Soviet Union and the United States balanced each other more or less effectively for nearly forty years and that the bipolar structure of the international system explains much of that stability.[4] However, there are important, if not tragic, questions to be asked. Not only why did the wartime alliance collapse so dramatically as to lead to the point of war in Europe (the Berlin crisis), but why did the United States, in particular, continue to expand its commitments far beyond the strategic areas of Western Europe and the offshore East Asian periphery? In fact, the culmination of the "origins" of the U.S. policy of containment was the Korean War—not the initial intervention to contain the North Korean attack, but rather the attempted "rollback" of communism to the Yalu River which provoked the full-scale military intervention of China. Not only did this effort at rollback create an entirely new war lasting over two more years, but it reinforced ties between China and the Soviet Union against the United States. Ultimately, although not explored here, these events set the stage for a second costly East Asian intervention: the "Americanization" of the Vietnam War in 1965, a sustained buildup until 1968, and then a prolonged exit even after recognizing the impossibility of military victory over North Vietnam. Notably, early realists such as George Kennan and Hans Morgenthau were among the first to question these wars, but the puzzle remains for contemporary realist (or neorealist) theory: why did the U.S. engage in such extended and self-defeating wars beginning in Korea?[5]

Taken together, the above examples make a critical point: systemic explanations of war cannot alone account for the behavior of key great powers in the twentieth century's major conflicts. This statement poses a problem for contemporary realist theory. Not only are the origins of WWI, WWII, and the Cold

---

[4] Also, from a domestic politics and decision-making perspective, the U.S. and Soviet political systems were reasonably stable such that their crises (e.g., the Cuban missile crisis) were usually handled by relatively effective small-group decision structures (e.g., the National Security Council and the Politburo).

[5] Snyder (1991: ch. 7), Stein (1993), and Kupchan (1994: ch. 7) each analyzes U.S. foreign policy after WWII; their analyses as well as that by Christensen (1996) linking the dynamics of the "origins" of the Soviet-U.S. Cold War to the expansion of U.S. commitments to East Asia and, ultimately, war against China are especially compelling, and can be extended to the similarly puzzling American commitment to Vietnam (also see Gelb with Betts, 1979; Berman, 1982).

War major international events, but these situations represent precisely the sort of phenomenon that should be explained best by this theory—that is, how states respond to the threats posed by an international system on the brink of war. Once again, the premise here is that decision-making theory can provide additional insights into these (and other) puzzling episodes. At the same time, though, the reader should keep in mind that what follows is certainly not an argument that decision-making structures somehow created or singularly drove these international conflicts. The historical research examined below makes clear that there were very real pressures from the international system (as well as domestic systems), and that leaders in these powers were more or less rationally attempting to cope with these clearly dangerous situations. What decision-making analysis can usefully explain is why leaders responded in ways that seem distorted in terms of systemic imperatives and their states' military capabilities. In other words, in line with Rosecrance's (1995) "Goldilocks" problem, the puzzle is why do states overreact or underreact to international pressures—or, following Richard Snyder and his associates (1954), how do they cope with the international situation as they define it.

## Evidence That War Decisions Are Complex

This section's survey of historical evidence is intended to show that decision-making situations prior to WWI, WWII, and the Cold War were quite complex and, thus, the responses to international threats by the great power leaders were not entirely obvious. In other words, systemic assumptions about decision making did not hold, which arguably accounts for some of the state responses that pose puzzles to realist theory as discussed above. As explicated by Waltz (1979; see also Bueno de Mesquita, 1981), systemic theories assume that decision makers respond more or less directly to the systemic imperatives posed by an anarchic international order. Particularly in international crises, the premise is that the dangers of war are so clear-cut that decision makers recognize the threat and can quickly agree on strategies to deal with it, focus exclusively on the goal of national security, and have the foreign policy authority necessary to commit the state's resources in responding to the threat. In short, systemic explanations of foreign policy apply under three conditions: information certainty, goal maximization, and the presence of an essentially unitary actor.

When looking at the twentieth century's great conflicts, however, these assumed decision-making conditions simply do not hold up very well. As a result, the responses of decision makers to even the most threatening situations often did not make sense in terms of international system imperatives. An analysis of these conflicts suggests that there are at least three conditions that make foreign policy problems (in our case, the threat of war) fundamentally more "complex" than is generally assumed in systemic theory:

> First, there is uncertainty, i.e., imperfect correspondence between information and the environment. Second, two or more values are affected by the decision, and there is a trade-off relationship between the values such that a great return to one can be obtained only at a loss to the other. Third, the power to make the decision is dispersed over a number of individual actors and/or organizational units. (Steinbruner, 1974:16)

The value of this concept of "complex decision" is that it isolates precisely those empirical conditions that are likely to make decision-making responses to systemic imperatives more problematic—or, at the least, not automatic or obvious. Where there is "uncertainty" about threats and how to deal with them, governments' responses will depend upon how leaders perceive and interpret the threats based on their own belief systems. When leaders confront "trade-offs," responses to threats involve simultaneous judgments about other policy issues, including how foreign policy actions will influence the government's hold on power (or vice versa). Moreover, in governments where political authority is dispersed, state actions revolve around the political maneuvers necessary to achieve agreement to support an alliance, defense expenditures, or ultimately the declaration of war. Furthermore, as will be argued at the close of this essay, the extent to which authority is dispersed will significantly affect how a government resolves the choices posed by uncertainty and value trade-offs.

The main point of this essay is simply that the decisions that were made leading to WWI, WWII, and the Cold War were fundamentally complex and involved uncertainty, value trade-offs, and the dispersion of authority. The argument is based on the wealth of historical research that has emerged on each of these conflicts over the past few decades. What follows is a concise survey of rather standard historical evidence indicating that complexity pervaded the situations in the approach to the First and Second World Wars and the origins of the Cold War. The reader should note that the evidence presented here, however, is not compiled from case study research by IR theorists such as Snyder and Kupchan; nor is it based upon studies by political scientists employing or advocating a decision-making perspective. Rather, this evidence is drawn from widely cited secondary analyses by historians on government decision making in the three conflicts. Three types of historical sources are used. The first are general historical studies of the origins of each conflict, particularly those that examine foreign policy decision making in the various belligerents and, as such, show the interactive sequence of state decisions that escalated the conflict. The second set of sources are country-specific historical accounts of each power's foreign policy decision making leading up to WWI, WWII, or the Cold War. These studies are often impressive, not only in terms of their rich explanation of policy debates, political structures, and specific decisions, but also in their review of historiographical debates concerning each power's entry into war. The third set of sources are general domestic political histories for each of the great powers. Although usually not focused on foreign policy and war, these studies survey the larger domestic context at the time, recounting contending political groups, competing policy issues, and larger political structures. In this way these latter sources place specific war decisions in the stream of domestic developments across the longer time frame associated with the buildups to WWI (e.g., 1890s–1914), WWII (1930s), and the Cold War (1945–1950).[6]

## *The Pervasiveness of Uncertainty Under Threat*

A key theme in this historical literature is that governments typically confronted significant uncertainty in responding to what now appear to be obvious threats in the periods leading up to WWI, WWII, and the Cold War. Often problematic was the choice between alternative strategies for coping with the threats and the assessment of the political and military risks involved in the actual use of military force. Although systemic arguments do not preclude debates within the government about threats and capabilities, the point here is that fundamental and often unresolvable uncertainty exists within the political leadership if there are opposing belief systems regarding what to do in foreign affairs.[7] For our purposes here, the key indicator of uncertainty will be the coexistence of "moderate" and "hard-line" mindsets among policymakers in the government regarding adversaries prior to the outbreak of war.[8] Moderates favor diplomatic accommodation and multilateral cooperation to avoid war, often because of an aversion to the risks (domestic or international) of going to war or, more extremely, because of pacifist or isolationist logics. Hard-liners,

---

[6] No claim is made to have consulted primary sources or even lengthy narratives such as those by Albertini (1952–57) or Watt (1989). The concern here is not to report original findings on these conflicts; rather, the goal is to judge IR theoretical assumptions in terms of the received wisdom among historians on the origins of these wars. The works cited are useful in that they synthesize the various, often contentious, strands of research on each country's role and motives in these conflicts. For a full discussion of the use of standard historical sources (as well as a comprehensive listing of them) in constructing political histories of the great powers since 1815 see Hagan (2000).

[7] In contrast, studies of perception and misperception get at uncertainty by looking at the judgments of decision makers as they responded to conditions as outside observers now know them. See, in particular, Levy's (1983) comparison of misperceptions before the First and Second World Wars as well as discussions in general works by Jervis (1976) and Lebow (1981).

[8] This distinction is based mainly on Vasquez (1993); see also Snyder and Diesing (1977). Although in another place the author (Hagan, 1994) has elaborated upon this distinction, the basic dichotomy is sufficient for identifying the existence of alternative policy positions. The main caution here is that hard-line and moderate positions vary across political systems as well as within political systems.

in contrast, call for sustained confrontation with adversaries including threatening the use of military force, which they view as the only viable way of deterring an inherently aggressive adversary—at the extreme are hypernationalist arguments favoring military aggression or expansion. Although, like any dichotomy, this hard-line–moderate distinction is simplistic, it still provides an efficient and theoretically meaningful device for making the point that quite often governments do not operate according to one well-defined strategy or clear systemic imperative. More important, the distinction adequately captures much of the rich historical research that suggests that, even at the brink of war, governments have to make choices under conditions of great uncertainty.

Although the pre-WWI period saw the rise of hard-liners in the governments of most of the great powers, these groups had not entirely supplanted the more moderate arguments that had restrained European affairs since the 1870s. A wide variety of views was especially evident in governments of the Triple Entente, where alternative arguments coexisted right up to the outbreak of WWI. The political leadership of Third Republic France had long been divided over the question of "normal" relations with Germany, with little opinion actually calling for a *revanchist* war to regain the lost provinces of Alsace and Lorraine. Rather, hard-liners called for assertive diplomacy against Germany, mainly through a strong alliance with Russia and ties with Britain, while moderates favored Franco-German détente via cooperation on trade and colonial questions. Although dating back to the defeat by Prussia, these divergent positions had, if anything, been sustained by the simultaneous revival of nationalism and socialism in prewar French politics.[9] As for autocratic Russia, the rising hard-line among the country's elites in the decade before the war was directed mainly toward Austria-Hungary over Balkans questions. The members of the Tsarist elite were less united about relations with Russia's other adversary—Germany. Hard-liners favored confronting Austria-Hungary while "deterring" Germany by building strong alliance ties with France and Britain, while moderates feared the cost of a premature war and advocated the "deflection" of German power through détente on common issues.[10] In Britain, the ruling Liberal Party, although viewed by the electorate as the "peace party" when compared with the Conser-

---

[9] For the variety of French viewpoints prior to WWI see Keiger (1983) and Hayne (1993). Several of the general French histories focus on the logic of competing hard-line and moderate beliefs through the nineteenth century, e.g., Zeldin (1973); Agulhon (1990); and Wright (1995). Tombs (1996) is especially useful in showing the strength of moderate, risk-averse arguments growing out of domestic upheaval (e.g., 1790s, 1848, 1871) and military defeat (1815 and 1871).

[10] These descriptions are by Leiven (1983), who emphasizes the diversity of elite opinion in Tsarist circles; but see also works by Geyer (1987), Spring (1988), McDonald (1992), and Neilson (1995).

vative Party, actually encompassed a wide range of foreign policy views once in power after 1905. These ranged from "liberal imperialists" favoring strengthened alliances with France and Russia (e.g., the foreign office under Lord Grey) to "radicals" with a Gladstonian aversion to realpolitik and ideological sympathy for human rights (e.g., in Russia). Even in August 1914 at the height of the crisis, Liberal leaders were divided over the necessity and/or morality of Britain's intervention in the war.[11]

Admittedly, the range of debate in Austria-Hungary and Germany (the two powers that started the war) was relatively narrow compared to that within the Entente powers.[12] In 1897 in a crucial reorganization of the imperial German government, William II appointed hard-liners to positions in the foreign ministry (Bulow) and Navy (Tirpitz). Along with the Kaiser, these leaders advocated a militant diplomacy to break Franco-Russian encirclement, a further tightening of the alliance with Austria-Hungary, and the pursuit of *Weltpolitik* in the form of colonial expansion and naval buildup.[13] A similar consolidation of hard-line dominance occurred in Austria-Hungary in 1906, with hard-liners appointed to the Foreign Ministry (Aehrenthal; later Berchtold) and army leadership (Conrad von Hotzendorf). These individuals believed that only a more assertive foreign policy in the Balkans would enable the empire to escape its rising ethnic tensions.[14] Yet all this change and consolidation did not preclude meaningful debate, particularly when it was recognized that the hard-line policies could increase, not decrease, the vulnerability of the two empires. The weakness of Austria-Hungary's situation posed uncertainties about the merits of war as a means of arresting the empire's decline, and thus hard-liners were

---

[11] Political divisions among the British leaders are documented in works by Williamson (1969), Steiner (1977), Kennedy (1980, 1981), Brock (1988), Chamberlain (1988), Wilson (1995), Young (1997), and Ferguson (1999).

[12] Note that just the opposite was the case for Italy, the third member of the Triple Alliance, whose choices were severely complicated by alternative adversaries (France or Austria-Hungary) and military defeat (Adowa). On prewar Italian foreign policy views see Thayer (1964), C. Seton-Watson (1967), Bosworth (1983), Chabod (1996), Clark (1996), and Mack Smith (1997).

[13] Most German histories acknowledge a fundamental shift in the course of the Second Reich with the rise of its pre-WWI hard-liners and that these changes were especially important on the road to WWI. See Berghahn (1973), Kennedy (1980), and Mommsen (1995) as well as general histories such as those by Ramm (1967), Holborn (1969), and Craig (1978). Kagan's (1995) analysis of WWI's origins begins with the fall of Bismarck and then addresses these subsequent leadership changes.

[14] Sked (1989), Williamson (1991), Fellner (1995), and in the larger context, Okey (2001) note the importance of this leadership change, while acknowledging the persistence of some dissenting arguments up through WWI.

constrained by less risk prone officials, such as the aging Emperor Franz Josef, the reformist Archduke Ferdinand, and the Hungarian Prime Minister Tisza. Nor was the more powerful Germany immune from policy questions, given its encirclement by the growing power and cohesion of the Triple Alliance. Thus, despite the talk of preventive war among some leaders (e.g., the 1912 "war council"), a sort of "preventive diplomacy" was pursued by the Kaiser to attract his Tsarist counterpart, while his chancellor (Bethmann-Hollweg) held out hope for relaxed tensions with Britain and, failing that, at least the country's neutrality in the case of war.[15] By July 1914, even though the Archduke Ferdinand's assassination provided an excellent opportunity to take preventive action against the Serbian threat and Entente encirclement, throughout the crisis advocates of preventive war in both governments had to fend off, first, the diplomatic arguments against war by the Hungarian prime minister and, later, the risk-averse hesitations of the German Kaiser and his chancellor about the escalating crisis.

The range of debate in the major powers prior to World War II is even more dramatic. At one extreme, Nazi Germany and militarist Japan manifested relatively little debate. Both these countries had foreign policies geared to systematic expansion through the use of military force.[16] Hitler's Germany comes closest to having had a grand plan of expansion with the goals of dismantling the Versailles settlement, gaining diplomatic dominance in the West, and subjugating the Slavic states to the East. Japanese foreign policy under the militarists is arguably not nearly as coherent, but the range of *policy* debate was relatively limited, having to do with the pace and direction of Japan's expan-

---

[15] Sources detailing prewar debates, and especially the position of Bethmann-Hollweg in Wilhelmine Germany include Stern (1967), Berghahn (1973), Craig (1978), Kennedy (1980), Schmidt (1990), and Herwig (1997). Of course, other research in the Fischer tradition stresses the coherence of the German hard-line; along with Fischer (1967), see Kaiser (1983), Wehler (1985), Pogge von Strandmann (1988), and Röhl (1995).

[16] The expansionist core of Nazi foreign policy is widely noted, although up until the latter part of the 1930s there were different views over the pace of that expansion; see works by Holborn (1969), Hildebrand (1970), Craig (1978), Michalka (1983), Müller (1983), Bell (1986), Berghahn (1987), Kaiser (1992), and Kagan (1995). On Japanese policy debates see works by Borton (1970), Fairbank, Reischauer, and Craig (1973), Barnhart (1987, 1995), Beasley (1987), Sagan (1988), Iriye (1997), and Jansen (2000). The foreign policies of Stalin's Soviet Union and Mussolini's Italy are typically portrayed as far more reactive and hardly conforming to a systematic plan. But to the extent that uncertainty existed, it was in the mind of each leader as neither faced competing policy arguments in the 1930s. On Mussolini see, e.g., Bosworth (1996), Mack Smith (1997), and Knox (2000). On the USSR see McCauley (1993), Westwood (1993), and Hosking (2001).

sion in China and the Pacific. Far more striking than the absence of broad debate in these authoritarian regimes is the narrow range of debate in one key democracy—Great Britain. Throughout the interwar period, British foreign affairs revolved around a consensus in the leadership that, first, recognized some legitimacy in German complaints about the Versailles settlement and, second, avoided any renewed commitment to the defense of France. This consensus was reinforced by the growing German threat after 1935 and, indeed, culminated at Munich in a shift from "passive" to "active" appeasement with Chamberlain's aggressive diplomacy.[17] Only after Hitler's unlimited goals became apparent with the dismantling of non-German Czechoslovakia did this policy raise significant doubts. But, remarkably, as Lukacs (1999) has recently documented, this stark policy failure did not create a solid hard-line consensus. Rather, even at the time of France's collapse in May 1940, the British cabinet remained divided over seeking a comprehensive peace with Hitler via Mussolini (e.g., Halifax) vs. continuing the war (e.g., Churchill).

Like this final British debate, increasing division was represented in the responses of France and the United States to the mounting threats after the mid-1930s. Both of these polities had debates over the most basic issues of how—and, indeed, whether or not—to respond to foreign aggression. Throughout the interwar period French political leaders (usually in multifaction/party cabinets) exhibited a clear "ambivalence" over the choice between competing hard-line and moderate strategies for dealing with the inevitable revival of German power.[18] Indeed, by the mid-1930s (e.g., the Spanish Civil War) right-wing and left-wing blocs were polarized over France's choice of allies and, ultimately, the question of which was worse—coexisting in an authoritarian Europe under Germany or in a communist order led by the Soviet Union. In the United States, the rising German and Japanese threats also provoked polarizing debates. Although the Roosevelt administration gradually came to recognize the imperative of deterring foreign aggression, isolationist arguments remained a very powerful force throughout American politics and particularly within Congress. Even after the fall of France, no consensus existed among

---

[17] These versions of appeasement are from Adams (1993), a source that offers a concise overview of the limited range of debate over appeasement. See also works by Bell (1986), Hughes (1988), Kagan (1995), and Young (1997). General political histories note the broad consensus within the interwar Conservative Party, e.g., Blake (1970), Beloff (1984), Williams and Ramsden (1990), Lloyd (1993), and Robbins (1994).

[18] Adamthwaite (1995) and Young (1996) are especially useful in portraying the politics underlying French foreign policy during the interwar years. Also see Azema (1984), Bernard and Dubief (1975), Agulhon (1990), McMillan (1992), and Larkin (1997).

U.S. leaders for responding to Churchill's pleas for military and economic aid for Britain's defense against the impending Nazi onslaught. Only Japan's attack on Pearl Harbor resolved American uncertainty regarding the relevance of the German threat in Europe and Japan's conquests in East Asia to U.S. security.[19]

Turning to the Cold War, it would be expected that the level of uncertainty in a bipolar world would be fundamentally less when compared to the previously discussed multipolar systems (e.g., Waltz, 1979). To an extent this is true, but there are still some interesting complications. Postrevisionist research on the "origins of the Cold War" stresses the uncertainties in the rise of Soviet-American tensions after WWII. Once again, the Soviet case under Stalin arguably revolved around his personal assessment of threats. Although acknowledging that Stalin had definite postwar goals and demands, key literature shows that Soviet postwar actions were not guided by a grand plan and that, instead, his actions were often ad hoc.[20] Far more striking is the American case, particularly if one recognizes that it took half a decade for hard-line containment policies to take shape. Initially, the Truman administration attempted to sort out Soviet moves in Eastern Europe and the Near East using competing mindsets—what Yergin (1977) calls the moderate "Yalta" axioms favoring continued normal diplomacy vs. the more hard-line "Riga" axioms pushing for confrontational diplomacy toward Stalin. Subsequently, after having defined the Soviet government as *the* dominant postwar threat, new uncertainties arose over the nature of that threat: was it essentially political and economic (Kennan) or was it essentially military (Acheson and Nitze)?[21] The latter view won out (in NSC-68), but still new uncertainties appeared when the "loss of China" revived debates about the geographical range of the communist threat; that is, was it Western Europe or East Asia, or both? Only the shock of the North Korean invasion of South Korea created, as had Pearl Harbor nine years earlier, a firm domestic consensus about U.S. international commitments.[22] So certain was this new consensus that the U.S. leaders became trapped by arguments insisting

---

[19] On interwar policy divisions in the United States see Divine (1965), Dallek (1979, 1983), and Small (1996) as well as general treatments in Paterson, Clifford, and Hagan (1983), LaFeber (1989), and Schulzinger (1998).

[20] Mastny (1979) and Taubman (1982) are especially useful in conveying the uncertainty in Stalin's foreign policy, although of course such was not manifest in postwar Kremlin debates.

[21] Works by Gaddis (1978, 1982), Larson (1985), Leffler (1992), and Yergin (1997) are especially effective in portraying the uncertainty in postwar U.S. foreign policy and the progression of debates that shaped the evolution of these policies. On alternative Cold War postures see also Dallek (1983).

[22] See Jervis (1980) and LaFeber (1989) on the impact of the Korean War.

on intervention in the Third World periphery—first, the expansion of the Korean War to the north and ultimately against China, and, second, the Americanization of the Vietnam War in the mid-1960s. For the next twenty years this Cold War consensus meant that there was minimal uncertainty in the American view of world affairs; debate was restricted to hard-line options about alternative instruments of military containment.

## *Value Trade-offs Under Threat*

A second assumption found in systemic theories is that security threats override other policy concerns, both at home and abroad. This view is challenged by the historical research on the origins of WWI, WWII, and the Cold War which typically portrays leaders as dealing with at least three types of trade-offs. One set involved the trade-offs between the continental balance of power and imperial goals in Africa and Asia. These trade-offs affected how the major powers balanced because scarce resources for the defense of the "core" were often allocated to the protection of the "peripheries."[23] A second type of trade-off centered around balancing domestic vs. foreign policy priorities. In the 1930s a dominant reality was that the economic collapse of the Great Depression had severely undercut the financial resources available for rearmament, at least for those governments that adhered to conservative fiscal policies—most clearly Britain under the Conservative Party and France under Radical Party influence. Even in the more prosperous pre-WWI era, the resources available to European governments (both democratic and authoritarian) were constrained not just by Liberal reformers as in Britain, but also by archaic systems of taxation and military deployment defended by traditional elements such as the Prussian aristocracy.[24] The third type of trade-off concerned the larger domestic political context and, in particular, the desire of leaders to retain power and preserve their regimes. Without reverting to arguments about the primacy of internal politics, the evidence suggests that domestic politics conditioned how leaders responded to foreign pressures leading up to WWI, WWII, and the Cold War. In short, despite the immediacy of international threats, decisions to go to war often involved the logic of Putnam's (1988) "two-level game." Given the per-

---

[23] Kupchan (1994) emphasizes this trade-off. Such trade-offs raised further uncertainties because continental adversaries were potential allies in imperial matters and the other way around; e.g., although the British might ally with France and Russia in response to Germany, they were traditionally opposed to these two powers in colonial matters.

[24] In Britain this opposition was broken by the Liberals, but in France and Germany entrenched groups resisting social and political reform restricted the revenue available to the military (see, e.g., Lamborn, 1991, and D'Lugo and Rogowski, 1993).

vasiveness of these domestic political influences as well as the extensive research regarding their effects, it is worth focusing on this third type of trade-off in detail.

Mounting domestic political crises were a common problem for European rulers prior to WWI, being almost as disruptive as the deterioration of the European balance of power. This domestic political decay took place in two stages. In the first, conservative leaders in all five powers coped with escalating, domestic political crises in the decade prior to the war.[25] Not only was Tsarist Russia shaken by the 1905 revolution and Austria-Hungary's progressive ethnic disintegration, but leaders in all of the powers faced problems in managing tenuous majorities in legislative bodies; that is, Austria-Hungarian parliamentary deadlocks and calls for reform of the Dual Monarchy, the rising power of the Social Democratic Party in Germany in an increasingly polarized Reichstag, Russian nationalist and Panslavic extremists in the new Duma, French Socialist Parties encroaching upon Radical Republican dominance, and the flux posed by the emerging Labour Party in Britain and the civil unrest in Ireland.[26] In the second stage, this political decay was intensified as rulers shifted to diversionary political strategies in which foreign policy prestige and nationalism were used to mobilize domestic support and isolate internal oppositions.[27] One key turning point was Wilhelmine Germany's 1897 adoption of the policies of *Sammlungpolitik* and *Weltpolitik* to deflect socialist opposition. This strategy was not unique. Similar shifts occurred in Austria-Hungary in 1906, in Russia under the pressure of the new Duma monarchy after 1906, and with France's nationalist

---

[25] Useful overviews of these crises are found in Mayer (1969), Joll (1984: ch. 5), and Kagan (1995), while Kaiser (1990) and Levy (1990–91) place these crises into the larger domestic and international contexts.

[26] The rise of domestic oppositions and resulting political tensions are documented in the larger context in the standard nineteenth-century histories of each of the powers. For Austria-Hungary see May (1951), Macartney (1968), and Kann (1974), and assessments by Sked (1989), Williamson (1991), and Mason (1997); for Germany see Ramm (1967), Holborn (1969), and Craig (1978), as well as structural arguments in Wehler (1985) and qualifications in Berghahn (1973), Kaiser (1983), Herwig (1992), Retallack (1992), and Mommsen (1995); for Russia see works by H. Seton-Watson (1967), Rogger (1983), Westwood (1993), and Hosking (2001), as well as the structural analysis by Geyer (1987); for France see works by Bury (1956), Brogan (1957), Cobban (1965), Mayeur and Reberioux (1984), and McMillan (1992), and analyses by Anderson (1977), Gildea (1996), and Tombs (1996); for Britain see Feuchtwanger (1985), McCord (1991), Robbins (1994), Rubinstein (1998), and Pugh (1999), and especially Williams and Ramsden (1990).

[27] On diversionary approaches to foreign policy, see the critical analyses by Levy (1988, 1989), while Hagan (1994) explicates the alternative strategies from which leaders can choose in dealing with domestic opposition.

revival under Poincaré after 1911. Only in the case of Britain (and also Italy) did leaders not shift to diversionary strategies.[28]

In the July 1914 crisis, domestic pressures were pervasive, but they played out in different ways across the five powers—and hardly reflected the manipulative diversionary strategies originally adopted by most governments. The linkage was greatest in the Austro-Hungarian case because domestic and foreign policy were largely inseparable; that is, limited war against Serbia (even at the risk of a war with Russia) was driven mainly by the rising position of Slavs within this multinational empire. Domestic political pressures were also profound for Germany and Russia, though more in terms of avoiding domestic losses than achieving political gains. Pivotal decisions by William II (approving the "blank check") and Nicholas II (approving mobilization) were motivated, in part, by the fear that domestic audiences would not tolerate another backing down in a major crisis. These leaders, both of whom had urged restraint in prior crises, were now trapped by the "blowback" (Snyder, 1991) of their own diversionary strategies to the point of risking a dangerous war.[29] In the French and British cases, the dynamics of domestic political influences were

---

[28] The early adoption of diversionary strategies is covered in further detail in the following sources. On Germany, see works by Fischer (1967), Geiss (1972), Berghahn (1973), Kennedy (1980), Kaiser (1983), Wehler (1985), and Mommsen (1995). For Austria-Hungary, Bled (1987), Sked (1989), and Williamson (1991) argue that part of the post-1906 hard-line was an attempt to use foreign policy prestige to unify the monarchy. In Russia, the post-1906 ministers in the Duma monarchy, more than the Tsar, reverted to the use of diversionary strategies (e.g., Stoylpin); see discussions by Leiven (1983, 1993), Geyer (1987), and McDonald (1992). Among French historians, Poincaré is largely pictured as playing the nationalism card after 1912, although what was most notable here was that the government did not back down in the crisis for fear of provoking unrest; see especially Keiger (1983), McMillan (1992), and Hayne (1993). In the British case, Steiner (1977) is explicit in rejecting the diversionary arguments, but see also Gordon (1974), Kennedy (1980), and Chamberlain (1988). Also note that Italian governments under Giolitti's influence had also turned away from the highly nationalistic foreign policies of earlier governments (see Thayer, 1964; C. Seton-Watson, 1967; Bosworth, 1983; Clark, 1996; and Mack Smith, 1997). Interestingly, the prewar British and Italian governments were both relatively democratic and had fought costly colonial wars in southern Africa and Ethiopia.

[29] Comparative judgments are, of course, difficult to make, but general analyses of the July crisis typically portray these two leaders as clearly motivated by fears of the domestic consequences of backing down (see Joll, 1984; Rich, 1992; Kagan, 1995). On William II's political fears behind the blank check see, in particular, the chapters by Mommsen (1995) as well as Berghahn (1973) and Kennedy (1980)–note also that Bethmann-Hollweg was very pessimistic about the domestic consequences of going to war. The tone is similar in analyses of Russian decision making (Leiven, 1983, 1993; Geyer, 1987; McDonald, 1992; Hosking, 2001); Nicholas II ultimately conceded to mobilization when advisers raised the question of domestic reaction to backing down in the crisis.

more complex. Ironically, the actions of the otherwise politically unstable French Third Republic appear to have been the least driven by domestic politics; for example, Poincaré's necessarily reactive crisis management focused on maintaining the Russian alliance. Actually, the main effect of domestic politics (i.e., another parliamentary scandal, the Caillaux crisis) was to distract public and parliamentary opponents from what was happening until it was quite evident that Germany would attack. Britain's politically constrained Liberal Party (unlike its Conservative predecessor) did not—in fact, could not—engage in diversionary strategies in July 1914. Not only was the cabinet preoccupied with the threat of civil war in Ireland, but any public hint of continental war would have provoked a parliamentary uproar among the party's moderate mainstream and certainly its antiwar Radicals and Labour Party allies.[30]

Linkages between domestic politics and foreign policy are even more apparent in the case of WWII. Compared to conditions prior to the outbreak of WWI, the presence of opposition varied widely across the different powers. In the 1930s the leaders of some powers actually faced very little organized opposition and had little reason to expect domestic criticism of their conduct of foreign policy. Of course, in all of the totalitarian/authoritarian regimes (most notably Germany and the Soviet Union) opposition groups and autonomous political institutions such as legislatures that were important arenas prior to WWI had been largely neutralized and most publics were, if anything, enthusiastic about nationalistic foreign policies.[31] Political constraints in democratic Britain were not dramatically different, as the long-dominant Conservative Party continued to rely on a large parliamentary majority. Except for isolated critics such as Churchill, the Baldwin and Chamberlain cabinets encountered minimal opposition from a unified and disciplined party that shared a broad consensus in favor of appeasement.[32] Only in the American and French cases were there

---

[30] Along with Joll (1984) and Kagan (1995), domestic political considerations in Poincaré's handling of the July crisis are discussed in Keiger (1983) and (1995), and assessed in Magraw (1983), Cobb (1988), and Adamthwaite (1995). On the politics within British decision making see Gordon (1974), Steiner (1977), Kennedy (1980), and Ferguson (1999).

[31] The progressive, and thorough, suppression of wider opposition is covered in standard histories of Germany (e.g., Holborn, 1969; Craig, 1978; Hildebrand, 1984; Carr, 1991; also Kaiser, 1992) and the Soviet Union (e.g., McCauley, 1993; Westwood, 1993; Hosking, 2001). The suppression of parliamentary democracy and traditional institutions was not as complete but still present in Italy (e.g., Duggan, 1994; Clark, 1996; Mack Smith, 1997; and Knox, 2000) and especially in Japan (e.g., Fairbank, Reischaur, and Craig, 1973; Duus, 1976; Berger, 1988; Pyle, 1996).

[32] On the political dominance of the cohesive Conservative Party in interwar British politics see works by Beloff (1984), Williams and Ramsden (1990), Lloyd (1993), and

significant domestic constraints, and they were quite strong. American isolationist sentiment remained entrenched in the Congress, and Roosevelt was entering the 1940 election cycle at precisely the peak of the European crisis. Foreign policy was a potentially explosive issue; for example, open assistance to a desperate Britain could raise public fears that military aid would eventually draw the U.S. into another European war.[33] The political situation in France was also desperate. The depression and international politics progressively polarized French party politics, stimulating the rise of the socialist/communist Left and the increasingly authoritarian Right—at the expense of the factionalized Radical center. Unlike the underlying continuity of the Third Republic ministries before WWI, distinctively rightist and leftist blocs now alternated in power, neither of which was capable of governing for a sustained period with a cohesive parliamentary base.[34]

The WWII case also illustrates the alternative ways leaders can respond to domestic opposition. Diversionary strategies, generally prevalent prior to WWI, were not the dominant political dynamic in the 1930s. The governments facing the most severe oppositions—France and the United States—clearly did not attempt to use foreign policy as a means of dealing with domestic problems. If anything, they consciously did the reverse, avoiding the controversies expected from any course of action that involved deterring aggressive states or entering into controversial alliances (e.g., one with the Soviet Union).[35] At the other extreme, the need for foreign policy success and nationalism were far more apparent in the authoritarian/totalitarian states. But even here there is complexity. As a way of dealing with opposition, it is a stretch to say that diversionary strategies influenced the shape of Nazi and Communist foreign policy. The leaders of both these regimes had so thoroughly eliminated any opposition that they faced minimal constraints in carrying out expansionist plans (in Hitler's case) and coping with difficult alignment choices (in Stalin's case). The cases of Japan and Italy, however, are more fluid, and it is here that diversionary

---

Robbins (1994). Young (1997) and Hughes (1988) note the party's broad consensus on foreign policy, a point emphasized in Bell (1986) and Kagan (1995).

[33] Holsti's (1993) "destroyers for bases" case documents this potential explosiveness; on domestic opposition to the Roosevelt administration see not only LaFeber (1989), Guinsburg (1994), and Small (1996), but also Brogan (1985) and Jones (1995).

[34] On the instability of interwar French politics see Bernard and Dubief (1975), McMillan (1992), and Larkin (1997). Adamthwaite (1995) and Young (1996) view domestic politics as yet another area of painful choices for French leaders.

[35] Controversy avoidance was inherent in French ministerial politics (see sources in footnote 34). In the American case, much more depended on FDR's cautious and accommodating political approach; e.g., see works by Divine (1965), Dallek (1979), and Farnham (1997) as well as Haines (1981) and LaFeber (1989).

strategies arguably apply. In neither case had traditional institutions (e.g., the monarchy) been entirely eradicated; radical nationalism remained a potent force. Mussolini's shift to a more aggressive foreign policy in the mid-1930s was motivated in part by the need for a dramatic policy success to enhance the prestige of the regime as domestic problems persisted.[36] Although less clear-cut, diversionary strategies in Japan's foreign policy reflect the attempt by contending factions to appeal to the nationalism of larger audiences as a way of discrediting their factional opponents.[37]

Domestic politics also conditioned the rise of Cold War tensions. Although rejecting the deterministic view that domestic structures alone propelled the U.S. and USSR confrontation, the postrevisionist literature suggests that the need for both Truman and Stalin to posture to domestic audiences contributed to the rise of hard-line policies. Even for the otherwise dominant Stalin, playing the ideological card of capitalist encirclement was a key means of reasserting dominance over the postwar Communist Party; for example, as evidenced by his hard-line reelection speech at the first postwar party congress in January 1946.[38] The prestige of a strong foreign policy would remain a fixture of Soviet foreign policy throughout the Cold War. The early political situation for the U.S. in the Truman administration was obviously different; yet, by democratic standards, this administration faced considerable opposition on its road to the Cold War.[39] Not only was the Republican Party bent on regaining its predepression/prewar dominance, the Democratic Party was by no means unified over Truman as a successor to FDR. Almost immediately after WWII (in the wake of the Yalta controversies), tough responses to puzzling Soviet actions enabled the vulnerable Truman to avoid accusations of appeasement, consolidate his control over the administration, and later mobilize public support for new Cold War commitments over the next few years. The Truman administration's ability to control anticommunist rhetoric, however, progressively declined, ultimately leading to extreme conservative backlash—or blowback—after the

---

[36] Duggan (1994), Bosworth (1996), Clark (1996), Mack Smith (1997), and Knox (2000), as well as Bell (1986) emphasize Mussolini's need for a foreign policy success to enhance the image of his regime.

[37] As explicated by Snyder (1991), public nationalism was crucial to the factional infighting within the Japanese government (see Ogata, 1964; Berger, 1988; Barnhart, 1995; Pyle, 1996).

[38] The need for Stalin to consolidate authority is evident not only in Cold War histories such as Mastny (1979) and Taubman (1982), but also in general histories such as those by McCauley (1993), Westwood (1993), and Hosking (2001).

[39] The larger domestic political scene is a prominent theme among Cold War historians, in particular, Gaddis (1972), Paterson (1979, 1988), Divine (1985), and LaFeber (1989); see also the analysis by Trout (1975).

"loss of China," severely weakening its credibility. As a result, the administration was hardly in a position to stand up to MacArthur's demands that communism be rolled back throughout Korea. These political lessons became an enduring fixture of the Cold War consensus. Indeed, the idea that foreign policy weakness could destroy an administration's credibility was a domestic certainty for U.S. presidents; for example, Kennedy's fear of impeachment in the Cuban missile crisis and Johnson's fears of political impotence if Vietnam collapsed.

## *The Diffusion of Political Authority Across Multiple Actors*

The final dimension in the concept of "complex decision" questions the premise that states act as unitary actors; that is, there is a contraction of authority and a single decision-making body evaluates distinct policy options and interests, sorts them out, and arrives at a final option. Historical research suggests that the key decisions surrounding WWI, WWII, and the early parts of the Cold War departed from this idealized pattern of decision making. At one extreme was the fragmentation of authority across competing factions, parties, or institutions, such that no single actor had the authority to commit the state to war, each wielding a veto over the others. Equally important, though, is the opposite extreme: the concentration of power into the hands of a single leader who either makes decisions alone or, if working with a single group, suppresses dissent so that viable alternatives are not seriously considered. The point here is that even the most severe international crises involve a wide variety of decision structures, including ones that operate in reasonable ways. Although effective single-group decision making undoubtedly occurs, the popular image of Kennedy's ExCom in the Cuban missile crisis is hardly typical of the crises considered here.

That there can be a variety of potential decision structures is quite apparent in the period prior to WWI and in the July 1914 crisis. Turning first to the powers that launched the war, neither Franz Josef nor William II dominated foreign policymaking, despite their personal constitutional authority over foreign policy and military affairs. In Austria-Hungary the ultimate decision-making body was the Crown Council, consisting of the emperor, the empire's common ministers (army, foreign affairs, and finance), and the Austrian and Hungarian prime ministers responsible to the separate parliaments in their respective halves of the empire. This decision-making body, an institution dating back to the 1867 creation of the Dual Monarchy, formalized the emperor's consultation with the key ministers and established the influence of the Hungarian leadership on the decisions of the Habsburg ruling elite. This process was critical during the July 1914 crisis; by then, the Hungarian prime minister (Tisza) was the only remaining advocate of diplomatic restraint and, because of his de facto veto power within the Crown Council, could block military action against Serbia.[40] In contrast to Austria-Hungary, the federal constitution of imperial Germany did not provide for a cabinet body to coordinate imperial decision making. Whereas Bismarck was personally able to coordinate decision making, William II never established clear-cut lines of authority over an increasingly complex German government. Foreign policymaking was largely distorted by this erratic decision-making pattern which became evident in the July 1914 crisis. William II personally extended the "blank check" to Austria-Hungary, and only afterwards sought the required approval of Chancellor Bethmann-Hollweg and met with senior foreign and military advisers. This lack of coordination appeared again at the height of the crisis. After the Russian mobilization Bethmann-Hollweg and William II separately sought restraint over the military option (now advocated by the army and finally pursued by Austria-Hungary), but the German government gave mixed diplomatic messages, leading even their one ally to ask "who is in charge in Berlin?"[41]

Despite the common threat from Germany, decision-making structures in Britain, Russia, and France were quite different, ranging from highly cohesive to very fragmented. Surprisingly, French decision making had become the most cohesive when actually faced with the prospect of war. Despite the Third Republic's long record of weak cabinets (occasionally collapsing under foreign threat), alarm over the German threat since the second Moroccan crisis (1911) permitted the emergence of strong presidential authority over foreign policy. Thus, in the July 1914 crisis, President Poincaré was able to insulate French diplomacy (supporting Russia) and restrain military preparations, not only from moderate and antiwar opposition in the new socialist-led cabinet, but also from repercussions from the cabinet's alarm over the Caillaux scandal in parlia-

---

[40] Austro-Hungarian decision making in the July crisis is discussed by Stone (1966), Jannen (1983), Williamson (1983, 1991), Sked (1989), Fellner (1995), and Herwig (1997) as well as emphasized by Joll (1984), Williamson (1988), and Kagan (1995). This decision making reflected the well-established, though fragmented, constitutional authority of the Dual Monarchy created in 1867. See overviews in works by May (1951), Macartney (1968), Jelavich (1987), Sked (1989), and Berenger (1997).

[41] German decision making in the July crisis, particularly its uncoordinated character, is documented in works by Berghahn (1973), Kaiser (1983), Schmidt (1990), Mommsen (1995), and Herwig (1997) and is emphasized in the general accounts in Joll (1984), Rich (1992), and Kagan (1995). For studies that downplay differences among German leaders and see imperial decision making in this crisis as much more cohesive check Fischer (1967, 1990) and Röhl (1995). The institutional fragmentation of imperial Germany (especially without Bismarck's strong leadership) is discussed in works by Ramm (1967), Holborn (1969), Berghahn (1973, 1994), Craig (1978), Kennedy (1980), and Wehler (1985).

ment.[42] Russian decision making was also reasonably coherent and, more than any of the powers, took place within a single group. As stipulated by the "unified government" reforms of the 1905 constitution, Czar Nicholas II did not act on his own (as before the Russo-Japanese War) but, instead, regularly met with his Council of Ministers. The discussions in these meetings appear to have been reasonably wide ranging; but with the prior removal of the main advocate for avoiding war (Kokovtsov), the July crisis finally produced a consensus not to back down again in the Balkans.[43] More problematic was British cabinet decision making under the divided Liberal Party. Representation of the full range of the party's factions was a political prerequisite for any Liberal cabinet. So long as British diplomacy was limited to verbal commitments and war plans, Lord Grey's foreign office could conduct Entente diplomacy without major constraints. However, the question of war required the full approval of the cabinet, and in the July crisis neither Grey nor the prime minister (Asquith) had the authority to commit Britain to the Entente powers. Taking such action on their own at the height of the crisis would have provoked the defection of the party's "radical" faction and ended nearly ten years of Liberal rule. Only the German invasion of neutral Belgium, in violation of international law and human rights, enabled enough Radical leaders to legitimize their support of a continental intervention.[44]

---

[42] Poincaré's dominance in the July 1914 crisis is developed (and defended) best by Keiger (1983, 1995), but it is a consistent theme elsewhere (Hayne, 1993; Adamthwaite, 1995) and is a key fixture in the general studies by Joll (1984) and Kagan (1995). While most histories of Third Republic France recognize the prewar shift in foreign affairs authority to the presidency, they also stress the underlying continuity in the country's prewar cabinets—particularly, the stabilizing effect of the Radical Party's pivotal position in successive cabinets. Along with Keiger (1983), see ministerial analyses in works by Anderson (1977), Magraw (1983), Mayeur and Reberioux (1984), McMillan (1992), and Gildea (1996).

[43] Russian decision making in the July crisis is covered in studies by Leiven (1983, 1993), Joll (1984), Spring (1988), Geyer (1987), McDonald (1992), Rich (1992), Kagan (1995), and Neilson (1995). These studies all make note that the July 1914 decision reflected changes in the makeup of the Council of Ministers; McDonald (1992) analyzes the institutional evolution of the "united government" that established this body. For the larger institutional evolution of the Duma monarchy see H. Seton-Watson (1967), Rogger (1983), Westwood (1993), and Hosking (2001).

[44] These constraints are well documented, not only in Joll (1984) and Kagan (1995), but also in studies of British decision making in the July crisis, e.g., Steiner (1977), Brock (1988), Wilson (1995), and Ferguson (1999). That these factional splits were not an aberration is evidenced in examinations of the evolution of the Liberal Party in Victorian Britain; see, e.g., Feuchtwanger (1985), Williams and Ramsden (1990), Parry (1993), Robbins (1994), and Rubinstein (1998).

Turning to the Second World War, the variety of decision structures is, like so much else about that war's origins, far more clear-cut when compared to 1914. However, by the mid-1930s, when the German and Japanese threats were apparent, decision making in none of the powers appears to fit the single-group model. There was, of course, the extreme concentration of authority in Fascist Italy, the Communist Soviet Union, and Nazi Germany. Not only did each regime have totalitarian arrangements in which governing authority was lodged entirely in a single institution, but Mussolini and certainly Hitler and Stalin had ruthlessly consolidated their own personal authority over foreign policy issues.[45] In all three cases decisions related to war reflected the unfettered power of a single individual. Far more striking is the one additional case: the highly cohesive decision making in the British cabinet under Prime Minister Chamberlain. Although structurally very different from the previous three regimes, British foreign policy authority was highly concentrated for a stable democracy. In part, this concentration was due to the dominance of interwar cabinets by a Conservative Party that, in contrast to the pre-WWI Liberal Party, was not factionalized and had a solid consensus on foreign policy issues (i.e., appeasement). All this was further intensified when Chamberlain became prime minister. More so than his predecessor (Stanley) and in response to mounting German demands, Chamberlain sought a comprehensive settlement through the "active" appeasement of Germany by aggressively conceding to their demands without war. However, in the process of doing so, he oversaw the resignation of those on the cabinet who were skeptics of these active concessions (e.g., Foreign Minister Eden) and by the time of the Munich crisis operated with minimal cabinet constraints.[46] Only when appeasement had failed (with Hitler's violation of the Munich accords) was Chamberlain's authority weakened, which (along with military failure in Norway) eventually led to his replacement by the more hard-line Churchill—a process of political adjustment that obviously did not occur in the totalitarian regimes.

France, the United States, and Japan all represent the fragmented extreme of pre-WWII decision making. In each, political authority was dispersed across politically autonomous factions, parties, and/or institutions. The French executive

---

[45] On Hitler's personal dominance of foreign policy, including his willingness to override the cautious advice of the military and diplomats, see Craig (1978), Bell (1986), Kaiser (1992), and Kagan (1995). Stalin's dominance over pre-WWII foreign policymaking is implied in political histories (e.g., McCauley, 1993) and noted in Bell (1986) and Kennedy-Pipe (1998). Despite the absence of totalitarian control, Mussolini did have control over foreign policy decisions, as argued in Clark (1996) and Mack Smith (1997) as well as Bell (1986).

[46] Chamberlain's dominance over the cabinet is noted in Adams (1993), Kagan (1995), Young (1997), and, more generally, Lloyd (1993).

responded to the rising Fascist threat with internal decay—not the greater coherence found before WWI. In contrast to the pre-WWI period, the cabinets of the interwar Third Republic retained close control over foreign policy. At no point did parliamentarians allow the emergence of strong presidential authority. These cabinets were, however, hardly in a position to fill the void. The centrist Radical Party lost its pivotal position in cabinets that now required greater power-sharing in coalitions with either the Left or the Right; furthermore, the opposing wings of the already loosely structured Radical Party were not in agreement over foreign policy. The result was that any cabinet, either center-left or center-right, faced collapse if it acted with conviction on foreign policy, and there was no institutional alternative to the cabinet.[47] The United States was also hampered by what were, in effect, coalition constraints. Although not approaching executive instability (or extra-parliamentary extremism), any meaningful commitments by an increasingly alarmed Roosevelt administration clearly required formal congressional approval. This constraint resulted from the earlier neutrality acts by which the largely isolationist Congress responded to international threats with increasing clamps on executive authority. Even after the fall of France (and actually beyond), Roosevelt had to cope with Congress's precise legal and political restrictions on aid to Great Britain. Only with Japan's late-1941 attack on Pearl Harbor was Congress fully willing to support FDR's control of foreign affairs.[48] Even Japan's actions—including its attack on the United States—were the manifestation of extremely fragmented decision making. Unlike its Italian and German allies, Japanese decision making under the militarist regime during the 1930s never achieved much coherence. Even after the demise of civilian influence via the Diet, policymaking authority remained ambiguously dispersed across highly competitive political factions tied to the army, navy, and imperial court. Although this fragmented decision making did not create the deadlock found in the U.S. and French cases, it surely helped distort Japan's judgments with respect to waging war throughout China and also across the Pacific against the U.S.[49]

The imperative of rising post-WWII tensions between the U.S. and the Soviet Union, like the previous decision-making conditions, would seem to have led

---

[47] The constraints posed by the divisions in the French cabinet are covered in works by Bernard and Dubief (1975), Azema (1984), McMillan (1992), Adamthwaite (1995), Young (1996), and Larkin (1997), as well as in Bell (1986) and Kagan (1995).

[48] On the assertion of congressional authority reflected in the neutrality legislation see Dallek (1979), LaFeber (1989), Guinsburg (1994), Small (1996), and Vieth (1996).

[49] On factional conflict in Japan's militarist regime see Ogata (1964), Fairbank, Reischauer, and Craig (1973), Hosoya (1974), Duus (1976), Fukui (1977), Berger (1988), Barnhart (1995), Pyle (1996), and Jansen (2000).

to more orderly decision structures in the Soviet Union and the United States. That was arguably often the case, even in the early period during the origins of the Cold War. There are few doubts that Stalin continued to make Soviet foreign policy decisions (as noted above), although the succeeding Khrushchev regime was never able to establish coherent foreign policymaking. Nor was American foreign policy decision making necessarily disorderly. President Truman had, in fact, returned White House decision making to the small-group norm; he consulted with his advisers in key crises over Poland, Berlin, and ultimately Korea. There is, though, one glaring exception. At key junctures, Truman administration decisions required congressional ratification and/or funding approval for new diplomatic and military commitments. Instead of conceding to isolationist opposition in the now Republican Congress, the Truman administration sought to mobilize its support by exaggerating the severity of the communist threat in both military and ideological terms.[50] The Truman doctrine speech in March 1947 was the primary manifestation of this manipulative strategy, one that within a few years would create difficulties in working with the Congress. Namely, the "blowback" of this rhetoric ultimately led to a "logroll" among the contending internationalist groups favoring European commitments and those emphasizing East Asia. This larger political framework not only enabled MacAuthur's expansion of the Korean War, but it ultimately necessitated the Americanization of the war in Vietnam by a Johnson administration fearful of losing congressional support for domestic programs and reforms. The hard-line Cold War consensus rested upon domestic political imperatives created after WWII.

## IMPLICATIONS OF THE HISTORICAL RECORD

The primary point of this article has been that decision-making conditions leading to the twentieth century's three great conflicts—WWI, WWII, and the Cold War—were fundamentally more complex than generally assumed by systemic explanations of war. This concluding section will use the historical evidence to suggest the importance of a decision units perspective on foreign policymaking which will be developed in the remainder of this special issue. The discussion here will be in two parts: (1) an examination of the assertion that the decision unit is a critical channel through which uncertainty and value trade-offs are defined—and that knowledge about the structure of the decision unit helps to explain the key puzzles about the origins of WWI, WWII, and the Cold War

---

[50] Along with Snyder (1991) and Christensen (1996) see analyses by Lowi (1967), Trout (1975), and Small (1996) and accounts of the origins of the Cold War by, for example, Gaddis (1972) and Paterson (1988).

described earlier in this essay; and (2) an exploration of the wide variety of types of decision structures and the fact that they vary independently of international systemic conditions or domestic regime structures–decision units are a theoretically fluid phenomenon that cannot be inferred directly from either systemic or domestic structures.

## Pivotal Role of Decision Units

One insight from the examination of the origins of WWI, WWII, and the Cold War is that decision units helped to define the degree of uncertainty and the value trade-offs in the situation. Even though wide-ranging debates occurred in most of the political systems, the structure of the decision unit governed the extent to which such debates were considered in the final decision to commit to war. For example, a predominant leader could exclude other positions, while a highly fragmented body enabled alternative positions and interests to coexist in a contradictory way or, worse, immobilize the government. Moreover, the structure of the decision unit affected how leaders responded to domestic political pressures. Decision structures could, on the one hand, position a national leader so that he/she could manipulate foreign policy politically and, on the other hand, magnify political conflicts within a political system thereby raising larger questions about the government's survival. Knowledge about the nature of the decision unit can help us account for the distortions in the balance of power in July 1914, at Munich, and in early Cold War crises—and, as such, can assist us in finding answers to the puzzles identified earlier in this article.

Learning about decision structures is key to understanding how the July 1914 crisis led to a war that none of the participants originally desired. With respect to the aggressive diplomacy of Austria-Hungary and Germany, perceived constraints on the decision unit of each negated their strategy of quickly suppressing Serbia and then forcing the Russians (and their French ally) to back down diplomatically under the threat of war. The structure of Austria-Hungary's Crown Council enabled the otherwise isolated opposition of the Hungarian prime minister to block the prowar consensus that now included even the emperor. During this lag, the Russians and the French became alarmed, and for the first time did not back down to the German threat of war. Decision structures were crucial in this new willingness to risk war with Germany. The Russian decision to mobilize is not entirely puzzling: its Tsar met with his Council of Ministers and considered options, which notably emphasized domestic consequences as much as international ones. Much more striking is the role of decision structures in the French case. Despite a cabinet dominated by moderates, if not antiwar socialists, the relatively hard-line French president (Poincaré) was able to control policy entirely, providing the Russian government with initial assurances and ultimately mobilizing the cabinet's full support for resisting Germany. In short, opposition to war was excluded from the decision process.

Decision units in the July 1914 crisis also undercut any deterrence against aggression. Most glaring is the failure of the Triple Entente to deter the German threat. This failure is mainly the collective fault of the British government (not just the foreign secretary, Lord Grey). Whereas Germany was prepared to fight Russia and France, they did not seek war with the British and assumed their neutrality. But because the British foreign policy leadership could not make a decision to go to war without consulting the entire cabinet—a cabinet dominated by opinion either unaware of, or outright opposed to, continental commitment—throughout the month of July they were unable to state openly to the Germans that they would side with the French and Russians in a war. Moreover, decision structures hindered attempts at mediation during this crisis by undercutting the impact of moderate arguments against hard-liners. In part, this effect resulted from the insertion of military leaders into the process, but it also stemmed from the weakness of equally alarmed civilian leaders. In the German case, the lack of a single decision-making body meant that hard-liners could undercut initiatives for restraint (e.g., those by Bethmann-Hollweg) by confronting the Kaiser separately and by re-framing any signals for restraining Austria-Hungary. Of course, any incentive for agreeing to the British foreign minister's proposal for an international conference was undercut by indications that Britain would remain neutral.

The role of decision structures is simpler in examining the road to the Second World War. In Germany, Britain, Italy, and Russia, strong leaders did not have to debate opposing positions (or could override them), nor were they forced to deal with the trade-offs across other issues. This control meant they were minimally constrained, and, as a result, could engage in risky, diplomatic initiatives. Thus, in the Munich crisis, Hitler could go to the brink of war and Chamberlain could grant the most extreme concessions without having to concede to skeptics. And, all the while, Mussolini and Stalin had the ability to engage in flexible diplomacy; for example, Stalin ultimately did not have constraints on dramatic alliance shifts. Nor did any of these leaders have to worry about domestic opposition to their moves; in fact, with the exception of Mussolini, their dominance largely negated the need to engage in diversionary approaches or to listen to advisers concerned with the possible domestic costs of any actions.

The situation in the more fragmented regimes—France, the United States, and Japan—was one that greatly amplified uncertainties and intensified domestic political constraints. For the U.S. and France, these constraints led to a failure of deterrence. The ambivalence in French responses to the rising German problem was embedded in fragile coalition governments, regardless of whether on the Left or the Right. In neither the center-left nor center-right coalition was there sufficient cabinet consensus to support an effective deterrent against Germany; for example, assertion of French military power or clear alignment choices with, say, Soviet Russia would have brought down the government. Nor could the United States then or after act as a deterrent. Despite the fact that FDR recognized the threat, the isolationist mindset had been written into the neutrality acts and other legislation—all with the effect that Congress could block any strong deterrent by the presidency. Furthermore, Congress represented a decision forum in which presidential assertion of U.S. power could be visibly challenged, thereby raising the electorate's fears that the U.S. was heading toward war. Similar to the French case, this possibility imposed severe constraints and led FDR's administration to engage in controversy avoidance. The Japanese case is, of course, just the opposite. Arguably more than any power its foreign policy was amplified by political dynamics within its decision structure. On the one hand, factional infighting magnified narrow policy differences and precluded any compromise or policy integration; on the other hand, factions openly played the nationalism card in order to mobilize public support for their interests.

The origins of the Cold War are also dominated by the contrast in decision structures between adversaries. Under Stalin, Soviet decision making remained highly concentrated. Although now able to operate with greater certainty than before the war, Stalin retained maximum flexibility in adjusting to the reassertion of U.S. power in world affairs. The only arguable difference was that he needed to consolidate authority at home, and thus played the nationalism—or ideology—card in defining anti-American hostility. The rise of American Cold War policies is far more complex and again rooted in domestic structures as the United States responded to a rising communist threat. As with FDR before WWII, the Truman administration had to gain a still isolationist/moderate Congress's formal approval for programs to implement containment via economic reconstruction and then deploy military forces. However, this time the result was not deadlock. Rather, the Truman administration systematically exaggerated the Soviet threat to mobilize support in Congress for authorizing containment commitments. This strategy ultimately resolved debate in the American public and became a fixture in U.S. Cold War policy—in fact, Congress accepted the most extreme view of the communist threat and ultimately trapped the Truman and successive administrations into an expanded array of East Asian commitments and two costly, self-defeating wars. Not until the painful lessons of Vietnam did the American political atmosphere again permit a reassertion of congressional authority and a wide-ranging debate over the premises of the Cold War and pragmatic engagement with the USSR.

## *Variability in Decision-Making Structures*

Another premise of this article and, in turn, this special issue is that decision structures are a highly variable phenomenon—all in ways compatible with Snyder and his associates' (1954) situationally grounded conception of decision making. The wide variation in types of decision structures described above is especially striking if one considers that we have been focusing on a rather narrow (albeit dramatic) aspect of international politics—the act of going to war. Even though the great powers were all responding to the same July 1914 crisis, the governments did not make decisions in the same way—one had a strong leader (France), others involved formal groups with some sharply divided (Austria-Hungary and Britain) and some not (Russia), and another (Germany) had decision-making authority that was especially fluid. The variation is even more clear-cut in the decisions made on the road to WWII: decision making typically involved either extremely cohesive states (Germany, Italy, the USSR, and even Britain) or highly fragmented states (France, the United States, and Japan). While small groups (e.g., the U.S.'s National Security Council and the Soviet Politburo) emerged as key Cold War decision-making bodies, it is still striking that the "origins" of that conflict involved a far wider array of decision structures. In short, contrary to theoretical expectations, the crises leading to WWI, WWII, and the Cold War, although handled by senior leaders, cannot be characterized as the result of a highly coherent decision process with a limited range of debate.

The range of variation in decision structures is also striking across time within each nation. The July 1914 crisis was one in a series of international crises, but this time the European powers did not back away from war—in part, because of changes in the makeup of their respective decision units. On the one hand, certain key leaders (William II, Franz Josef, and Nicholas II) now accepted war, while, on the other hand, key opponents to war in the previous crises were gone from the Russian and Austro-Hungarian decision-making bodies. In France, there had been an increasing concentration of authority in the presidency in the years leading up to 1914 so that decision making was dominated by a single leader capable of overcoming cabinet divisions. In Britain, a declaration of war required the formal approval of the full political leadership. That meant those in charge of foreign affairs had to deal with key opponents, announcing commitments unknown to the wider Liberal Party leadership and opposed by that party's Radical wing. World War II and the Cold War also present an interesting variation in decision structures, although they appear to be tied to regime types. Among the surviving democracies in 1940, policy failures led not only to policy adjustments but ultimately to a rearrangement of political authority. After the fall of France, the hard-line Churchill finally replaced Chamberlain as Britain's prime minister and the Roosevelt administration gradually gained greater authority to lend material support to the British. A decade later such a change occurred again in the United States: the rising Soviet threat contributed to a renewed dominance of the executive in making Cold War policy. These dynamics were just the opposite in the totalitarian regimes. Hitler's Germany, Mussolini's Italy, and Japan's militarists ultimately overextended themselves, in part because their leadership was politically entrenched and unwilling to adjust strategically. The record of the Soviet Union is more mixed from Stalin to the reformist Khrushchev and the consensus-oriented Brezhnev, but the failure to adjust was all too apparent by the late 1980s.

This last point does not, however, suggest that decision structures and their dynamics (including how such units respond to domestic pressures) are a direct function of internal regime structure—such as democracy and democratization. In fact, decision structures can vary even within the same type of regime structure. Thus, predominant leaders are found not only in totalitarian regimes (e.g., Nazi Germany, Soviet Russia, Fascist Italy), but also in democratic states such as France in 1914, Britain in the late 1930s, and the United States at the height of the Cold War. And fragmentation of authority occurs not just in established democratic systems, such as Britain in 1914, France in the late 1930s, and the U.S. in the late 1930s and mid-1940s, but also in authoritarian regimes such as Japan in the 1930s and in anocratic (or democratizing) regimes like Germany and Austria-Hungary in July 1914.

These observations lead to another point: the political dynamics of these decision structures are not the same. For example, consider Germany, Austria-Hungary, and Russia during the crisis leading up to WWI. Among just these three anocracies, there is little in common in decision making. Austro-Hungarian foreign policy is constrained by internal divisions, while that of imperial Germany is far more uncoordinated with few constraints and Russian foreign policy is being made by a single group. What these anocratic regimes had in common was how their leaders perceived vulnerability to domestic opposition: a domestic ethnic crisis propelled Austria-Hungary into war, and both William II and Nicholas II feared the domestic prospects of backing down in another crisis. Although relatively democratic, pre-WWI France and Britain also faced domestic crises, but they were either careful not to—or politically unable to—manipulate the crisis for domestic gain. And, like Britain in 1914, the leaders of neither the U.S. nor France in the 1930s were in a position to inflame international affairs to resolve domestic pressures—to project threats would have intensified domestic divisions, rather than quelled them. The dynamics were the opposite for militarist Japan and Mussolini's Italy. Regimes with a high concentration of authority were in a better position to manipulate foreign policy for domestic purposes, although again the pattern is not consistent—for example, only Mussolini seemed to be driven mainly by diversionary strategies. In the Cold War, diversionary strategies became more the norm—but for both the authoritarian Soviet Union and the democratic United States.

## In Conclusion

The basic point of this essay has been that decision making is an important (albeit fluid) phenomenon in international relations, even in the most severe of international conditions: that of the origins of war. As historical research has documented, how decision makers were configured had a significant impact on the outbreak of the twentieth century's great conflicts: WWI, WWII, and the Cold War. In each case, the decision units responsible for committing the resources of the government acted in ways that contributed to the conflict spiral, deterrence failure, or both. We are not arguing here that decision structures were the primary cause of these conflicts. Not only was the road to war a long one domestically and internationally, but the leadership in these countries was reacting to very real systemic pressures. Given, however, the complexity of these pressures (the degree of uncertainty they generated, the value trade-offs they provoked, and the dispersion of decision authority they encountered), it is not obvious how the leadership would (or should) respond. Furthermore, it should be evident from the previous discussion that the leadership's responses are not necessarily associated with domestic political structures. Not only are leaders rarely motivated solely by domestic concerns, but regime structures do no better than international imperatives in explaining decision structures and processes. Instead, decision units appear to operate in a way that is often independent of the otherwise compelling constraints of both international and domestic politics. As Richard Snyder's original formulation suggested and decades of subsequent historical research seems to demonstrate, decision-making dynamics cannot be inferred directly from international and domestic structures. Without denying the importance of both domestic and international situations, all that follows in this special issue is premised on the idea that decision units do matter, although in complex ways that vary across time and political systems.

## References

ADAMS, R. J. Q. (1993) *British Politics and Foreign Policy in the Age of Appeasement, 1935–39*. Stanford, CA: Stanford University Press.

ADAMTHWAITE, ANTHONY (1995) *Grandeur & Misery: France's Bid for Power in Europe, 1914–1940*. London: Arnold.

AGULHON, MAURICE (1990) *The French Republic, 1879–1992*. Oxford: Blackwell.

ALBERTINI, LUIGI (1952–57) *The Origins of the War of 1914*. London: Oxford University Press.

ALLISON, GRAHAM T. (1971) *Essence of Decision: Explaining the Cuban Missile Crisis*. Boston: Little, Brown.

ANDERSON, R. D. (1977) *France, 1870–1914: Politics and Society*. London: Routledge & Kegan Paul.

AZEMA, JEAN-PIERCE (1984) *From Munich to the Liberation, 1938–1944*. Cambridge: Cambridge University Press.

BARNHART, MICHAEL A. (1987) *Japan Prepares for Total War: The Search for Economic Security, 1919–1941*. Ithaca, NY: Cornell University Press.

BARNHART, MICHAEL A. (1995) *Japan and the World Since 1868*. London: Arnold.

BEASLEY, W. G. (1987) *Japanese Imperialism: 1894–1945*. Oxford: Clarendon Press.

BELL, P. M. H. (1986) *The Origins of the Second World War in Europe*. London: Longman.

BELOFF, MAX (1984) *Wars and Welfare: Britain 1914–1945*. Baltimore, MD: Edward Arnold.

BERENGER, JEAN (1997) *A History of the Habsburg Empire, 1700–1918*. New York: Longman.

BERGER, GORDON (1988) "Politics and Mobilization in Japan, 1931–1945." In *The Cambridge History of Japan, Volume 6: The Twentieth Century*, edited by Peter Duus. Cambridge: Cambridge University Press.

BERGHAHN, VOLKER R. (1973) *Germany and the Approach of War in 1914*. New York: St. Martin's Press.

BERGHAHN, VOLKER R. (1987) *Modern Germany: Society, Economy and Politics in the Twentieth Century*, 2nd ed. Cambridge: Cambridge University Press.

BERGHAHN, VOLKER R. (1994) *Imperial Germany, 1871–1914: Economy, Society, Culture, and Politics*. Providence, RI: Berghahn Books.

BERMAN, LARRY (1982) *Planning a Tragedy: The Americanization of the War in Vietnam*. New York: W. W. Norton.

BERNARD, PHILIPPE, AND HENRI DUBIEF (1975) *The Decline of the Third Republic, 1914–1938*. Cambridge: Cambridge University Press.

BLAKE, ROBERT (1970) *The Conservative Party from Peel to Churchill*. New York: St. Martin's Press.

BLED, JEAN-PAUL (1987) *Franz Joseph*. Cambridge, MA: Blackwell.

BORTEN, HUGH (1970) *Japan's Modern Century*, 2nd ed. New York: Roland Press.

BOSWORTH, R. J. B. (1983) *Italy and the Approach of the First World War*. New York: Macmillan Press.

BOSWORTH, R. J. B. (1996) *Italy and the Wider World, 1860–1960*. London: Routledge.

BROCK, MICHAEL (1988) "Britain Enters the War." In *The Coming of the First World War*, edited by R. J. W. Evans and Hartmut Pogge von Strandmann. Oxford: Oxford University Press.

BROGAN, D. W. (1957) *The Development of Modern France, 1870–1939.* London: Hamish Hamilton.

BROGAN, HUGH (1985) *The Penguin History of the United States of America.* London: Penguin.

BUENO DE MESQUITA, BRUCE (1981) *The War Trap.* New Haven, CT: Yale University Press.

BURY, J. P. T. (1956) *France 1814–1940.* London: Methuen.

CARR, WILLIAM (1991) *A History of Germany, 1815–1990.* London: Arnold.

CHABOD, FEDRICO (1996) *Italian Foreign Policy: The Statecraft of the Founders.* Princeton, NJ: Princeton University Press.

CHAMBERLAIN, MURIEL E. (1988) *"Pax Britannica" British Foreign Policy, 1789–1914.* London: Longman.

CHRISTENSEN, THOMAS J. (1996) *Useful Adversaries: Grand Strategy, Domestic Mobilization, and Sino-American Conflict, 1947–1958.* Princeton, NJ: Princeton University Press.

CLARK, MARTIN (1996) *Modern Italy*, 2nd ed. London: Longman.

COBB, RICHARD (1988) "France and the Coming of War." In *The Coming of the First World War*, edited by R. J. W. Evans and Hartmut Pogge von Strandmann. Oxford: Oxford University Press.

COBBAN, ALFRED (1965) *A History of Modern France, Volume 3: 1871–1962.* New York: Penguin.

CRAIG, GORDON A. (1978) *Germany: 1866–1945.* New York: Oxford University Press.

DALLEK, ROBERT (1979) *Franklin D. Roosevelt and American Foreign Policy, 1932–1945.* New York: Oxford University Press.

DALLEK, ROBERT (1983) *The American Style of Foreign Policy: Cultural Politics and Foreign Affairs.* New York: Alfred A. Knopf.

DIVINE, ROBERT A. (1965) *The Reluctant Belligerent: American Entry into World War II.* New York: John Wiley & Sons.

DIVINE, ROBERT A. (1985) *Since 1945: Politics and Diplomacy in Recent American History*, 3rd ed. New York: Alfred A. Knopf.

D'LUGO, DAVID, AND RONALD ROGOWSKI (1993) "The Anglo-German Naval Race and Comparative Constitutional 'Fitness'." In *The Domestic Bases of Grand Strategy*, edited by Richard Rosecrance and Arthur A. Stein. Ithaca, NY: Cornell University Press.

DUGGAN, CHRISTOPHER (1994) *A Concise History of Italy*. Cambridge: Cambridge University Press.

DUUS, PETER (1976) *The Rise of Modern Japan*. Boston: Houghton Mifflin.

FAIRBANK, JOHN K., EDWIN O. REISCHAUER, AND ALBERT M. CRAIG (1973) *East Asia: Tradition and Transformation*. Boston: Houghton Mifflin.

FARNHAM, BARBARA REARDEN (1997) *Roosevelt and the Munich Crisis: A Study of Political Decision Making*. Princeton, NJ: Princeton University Press.

FELLNER, FRITZ (1995) "Austria-Hungary." In *Decisions for War, 1914*, edited by Keith Wilson. New York: St. Martin's Press.

FERGUSON, NIALL (1999) *The Pity of War: Explaining World War I*. New York: Basic Books.

FEUCHTWANGER, E. J. (1985) *Democracy and Empire: Britain 1865–1914*. London: Arnold.

FISCHER, FRITZ (1967) *Germany's Aims in the First World War*. New York: Norton.

FISCHER, FRITZ (1990) "The Foreign Policy of Imperial Germany and the Outbreak of the First World War." In *Escape into War? The Foreign Policy of Imperial Germany*, edited by Gregor Schöllgen. Oxford: Berg.

FUKUI, HARUHIRO (1977) Foreign Policy by Improvisation: The Japanese Experience. *International Journal* **32**:791–812.

GADDIS, JOHN LEWIS (1972) *The United States and the Origins of the Cold War*. New York: Columbia University Press.

GADDIS, JOHN LEWIS (1978) *Russia, the Soviet Union, and the United States: An Interpretive History*. New York: Alfred A. Knopf.

GADDIS, JOHN LEWIS (1982) *Strategies of Containment: A Critical Appraisal of Postwar American National Security Policy*. New York: Oxford University Press.

GEISS, IMANUEL (1972) "Origins of the First World War." In *The Origins of the First World War: Great Power Rivalry and German War Aims*, edited by H. W. Koch. Hong Kong: Macmillan.

GELB, LESLIE H., WITH RICHARD K. BETTS (1979) *The Irony of Vietnam: The System Worked*. Washington, DC: Brookings Institution.

GEYER, DIETRICH (1987) *Russian Imperialism: The Interaction of Domestic and Foreign Policy, 1860–1914*. New Haven, CT: Yale University Press.

GILDEA, ROBERT (1996) *France 1870–1914*, 2nd ed. London: Longman.

GILPIN, ROBERT (1981) *War and Change in World Politics*. Cambridge: Cambridge University Press.

GORDON, MICHAEL R. (1974) Domestic Conflict and the Origins of the First World War: The British and the German Cases. *Journal of Modern History* **46**(June):191–226.

GUINSBURG, THOMAS N. (1994) "The Triumph of Isolationism." In *American Foreign Relations Reconsidered, 1890–1993*, edited by Gordon Martel. London: Routledge.

HAGAN, JOE D. (1994) Domestic Political Systems and War Proneness. *Mershon International Studies Review* **38**(supplement 2):183–207.

HAGAN, JOE D. (2000) Rulers, Oppositions, and Great Power Conflicts Since 1815. Manuscript, West Virginia University.

HAINES, GERALD K. (1981) "Roads to War: United States Foreign Policy, 1931–1941." In *American Foreign Relations: A Historiographical Review*, edited by Gerald K. Haines and J. Samuel Walker. Westport, CT: Greenwood Press.

HALPERIN, MORTON H. (1974) *Bureaucratic Politics and Foreign Policy*. Washington, DC: Brookings Institution.

HAYNE, M. B. (1993) *The French Foreign Office and the Origins of the First World War, 1898–1914*. Oxford: Clarendon Press.

HERMANN, CHARLES F., AND MARGARET G. HERMANN (1969) "An Attempt to Simulate the Outbreak of World War I." In *International Politics and Foreign Policy: A Reader in Research and Theory*, edited by James N. Rosenau. New York: Free Press.

HERWIG, HOLGER H. (1992) "Industry, Empire and the First World War." In *Modern Germany Reconsidered, 1870–1945*, edited by Gordon Martel. London: Routledge.

HERWIG, HOLGER H. (1997) *The First World War: Germany and Austria-Hungary, 1914–1918*. London: Arnold.

HILDEBRAND, KLAUS (1970) *The Foreign Policy of the Third Reich*. Berkeley: University of California Press.

HILDEBRAND, KLAUS (1984) *The Third Reich*. London: Allen & Unwin.

HOLBORN, HAJO (1969) *A History of Modern Germany, 1840–1945*. Princeton, NJ: Princeton University Press.

HOLSTI, OLE R. (1976) "Foreign Policy Decision Makers Viewed Psychologically: 'Cognitive Process' Approaches." In *In Search of Global Patterns*, edited by James N. Rosenau. New York: Free Press.

HOLSTI, OLE R. (1993) *The 1940 Destroyers for Bases Deal with Great Britain*. Washington, DC: Georgetown University Institute for the Study of Diplomacy.

HOSKING, GEOFFREY (2001) *Russia and the Russians: A History.* Cambridge, MA: The Belknap Press of Harvard University Press.

HOSOYA, CHIHIRO (1974) Characteristics of Foreign Policy Decision Making Systems in Japan. *World Politics* **29**:90–113.

HUGHES, JEFFREY L. (1988) "The Origins of World War II in Europe: British Deterrence Failure and German Expansionism." In *The Origin and Prevention of Major Wars,* edited by Robert I. Rotberg and Theodore J. Rabb. Cambridge: Cambridge University Press.

IRIYE, AKIRA (1997) *Japan & The Wider World: From the Mid-Nineteenth Century to the Present.* London: Longman.

JANIS, IRVING L. (1972) *Victims of Groupthink: A Psychological Study of Foreign Policy Decisions and Fiascoes.* Boston: Houghton Mifflin.

JANNEN, WILLIAM, JR. (1983) "The Austro-Hungarian Military Relations Before World War I." In *Essays on World War I: Origins and Prisoners of War,* edited by Samuel R. Williamson, Jr., and Peter Pastor. New York: Columbia University Press.

JANSEN, MARIUS B. (2000) *The Making of Modern Japan.* Cambridge, MA: Harvard University Press.

JELAVICH, BARBARA (1987) *Modern Austria: Empire and Republic, 1815–1986.* Cambridge: Cambridge University Press.

JERVIS, ROBERT (1976) *Perception and Misperception in International Politics.* Princeton, NJ: Princeton University Press.

JERVIS, ROBERT (1980) The Impact of the Korean War on the Cold War. *Journal of Conflict Resolution* **24**(December):563–592.

JOLL, JAMES (1984) *The Origins of the First World War.* London: Longman Group.

JONES, MALDWYN A. (1995) *The Limits of Liberty: American History, 1607–1992,* 2nd ed. Oxford: Oxford University Press.

KAGAN, DONALD (1995) *On the Origins of War and the Preservation of Peace.* New York: Doubleday.

KAISER, DAVID E. (1983) Germany and the Origins of the First World War. *Journal of Modern History* **55**(September):442–474.

KAISER, DAVID E. (1990) *Politics and War: European Conflict from Philip II to Hitler.* Cambridge, MA: Harvard University Press.

KAISER, DAVID E. (1992) "Hitler and the Coming of the War." In *Modern Germany Reconsidered, 1870–1945,* edited by Gordon Martel. London: Routledge.

KANN, ROBERT A. (1974) *A History of the Habsburg Empire, 1526–1918.* Berkeley: University of California Press.

KEIGER, JOHN F. V. (1983) *France and the Origins of the First World War.* New York: St. Martin's Press.

KEIGER, JOHN F. V. (1995) "France." In *Decisions for War, 1914,* edited by Keith Wilson. New York: St. Martin's Press.

KENNEDY, PAUL (1980) *The Rise of the Anglo-German Antagonism, 1860–1914.* Boston: Allen & Unwin.

KENNEDY, PAUL (1981) *The Realities Behind Diplomacy: Background Influences on British External Policy, 1865–1980.* London: Allen & Unwin.

KENNEDY-PIPE, CAROLINE (1998) *Russia and the World: 1917–1991.* London: Arnold.

KEOHANE, ROBERT (1984) *After Hegemony: Cooperation and Discord in the World Political Economy.* Princeton, NJ: Princeton University Press.

KNOX, MACGREGOR (2000) *Common Destiny: Dictatorship, Foreign Policy, and War in Fascist Italy and Nazi Germany.* Cambridge: Cambridge University Press.

KUPCHAN, CHARLES A. (1994) *The Vulnerability of Empire.* Ithaca, NY: Cornell University Press.

LAFEBER, WALTER (1989) *The American Age: United States Foreign Policy at Home and Abroad Since 1750.* New York: Norton.

LAMBORN, ALAN C. (1991) *The Price of Power: Risk and Foreign Policy in Britain, France, and Germany.* Boston: Unwin & Hyman.

LARKIN, MAURICE (1997) *France Since the Popular Front: Government and People, 1936–1996.* Oxford: Clarendon Press.

LARSON, DEBORAH WELCH (1985) *Origins of Containment: A Psychological Explanation.* Princeton, NJ: Princeton University Press.

LEBOW, RICHARD NED (1981) *Between Peace and War: The Nature of International Crisis.* Baltimore, MD: Johns Hopkins University Press.

LEFFLER, MELVYN P. (1992) *A Preponderance of Power: National Security, the Truman Administration, and the Cold War.* Stanford, CA: Stanford University Press.

LEIVEN, D. C. B. (1983) *Russia and the Origins of the First World War.* New York: St. Martin's Press.

LEIVEN, D. C. B. (1993) *Nicholas II: Twilight of the Empire.* New York: St. Martin's Press.

LEVY, JACK S. (1983) Misperception and the Causes of War: Theoretical Linkages and Analytical Problems. *World Politics* **35**:76–99.

LEVY, JACK S. (1988) "Domestic Politics and War." In *The Origin and Prevention of Major Wars*, edited by Robert I. Rotberg and Theodore K. Rabb. New York: Cambridge University Press.

LEVY, JACK S. (1989) "The Diversionary Theory of War: A Critique." In *Handbook of War Studies*, edited by Manus I. Midlarsky. Boston: Unwin & Hyman.

LEVY, JACK S. (1990–91) Preferences, Constraints, and Choices in July 1914. *International Security* **15**(winter):151–185.

LLOYD, T. O. (1993) *Empire, Welfare State, Europe: English History, 1906–1992*, 4th ed. Oxford: Oxford University Press.

LOWI, THEODORE J. (1967) "Making Democracy Safe for the World: National Politics and Foreign Policy." In *Domestic Sources of Foreign Policy*, edited by James N. Rosenau. New York: Free Press.

LUKACS, JOHN (1999) *Five Days in London, May 1940*. New Haven, CT: Yale University Press.

MACARTNEY, C. A. (1968) *The Habsburg Empire, 1790–1918*. New York: Macmillan.

MACK SMITH, DENIS (1997) *Modern Italy: A Political History*. Ann Arbor: University of Michigan Press.

MAGRAW, ROGER (1983) *France, 1815–1914*. New York: Oxford University Press.

MASON, JOHN W. (1997) *The Dissolution of the Austro-Hungarian Empire, 1867–1918*, 2nd ed. London: Longman.

MASTNY, VOJTECH (1979) *Russia's Road to the Cold War*. New York: Columbia University Press.

MAY, ARTHUR J. (1951) *The Habsburg Monarchy, 1867–1914*. Cambridge, MA: Harvard University Press.

MAYER, ARNO (1969) "Domestic Causes of the First World War." In *The Responsibility of Power*, edited by L. Krieger and F. Stern. Garden City, NY: Doubleday.

MAYEUR, JEAN-MARIE, AND MADELEINE REBERIOUX (1984) *The Third Republic from Its Origins to the Great War, 1871–1914*. Cambridge: Cambridge University Press.

MCCAULEY, MARTIN (1993) *The Soviet Union: 1917–1991*, 2nd ed. London: Longman.

MCCORD, NORMAN (1991) *British History, 1815–1906*. New York: Oxford University Press.

MCDONALD, DAVID MACLAREN (1992) *United Government and Foreign Policy in Russia, 1900–1914*. Cambridge, MA: Harvard University Press.

McMillan, James F. (1992) *Twentieth Century France: Politics and Society, 1898–1991.* London: Arnold.

Michalka, Wolfgang (1983) "Conflicts Within the German Leadership on the Objectives and Tactics of German Foreign Policy, 1933–9." In *The Fascist Challenge and the Policy of Appeasement*, edited by Wolfgang J. Mommsen and Lothar Kettenacker. London: Allen & Unwin.

Mommsen, Wolfgang J. (1995) *Imperial Germany, 1867–1918: Politics, Culture, and Society in an Authoritarian State.* London: Arnold.

Müller, Klaus-Jürgen (1983) "The German Military Opposition Before the Second World War." In *The Fascist Challenge and the Policy of Appeasement*, edited by Wolfgang J. Mommsen and Lothar Kettenacker. London: Allen & Unwin.

Neilson, Keith (1995) "Russia." In *Decisions for War, 1914*, edited by Keith Wilson. New York: St. Martin's Press.

Ogata, Sadako (1964) *Defiance in Manchuria: The Making of Japanese Foreign Policy, 1931–32.* Berkeley: University of California Press.

Okey, Robin (2001) *The Habsburg Monarchy.* New York: St. Martin's Press.

Parry, Jonathan (1993) *The Rise and Fall of Liberal Government in Victorian Britain.* New Haven, CT: Yale University Press.

Paterson, Thomas G. (1979) *On Every Front: The Making of the Cold War.* New York: Norton.

Paterson, Thomas G. (1988) *Meeting the Communist Threat: Truman to Reagan.* New York: Oxford University Press.

Paterson, Thomas G., J. Garry Clifford, and Kenneth J. Hagan (1983) *American Foreign Policy: A History Since 1900*, 2nd ed. Lexington, MA: D. C. Heath.

Pogge von Strandmann, Hartmut (1988) "Germany and the Coming of War." In *The Coming of the First World War*, edited by R. J. W. Evans and Hartmut Pogge von Strandmann. Oxford: Oxford University Press.

Pugh, Martin (1999) *Britain Since 1789: A Concise History.* New York: St. Martin's Press.

Putnam, Robert D. (1988) Diplomacy and Domestic Politics: The Logic of Two-Level Games. *International Organization* **42**:427–460.

Pyle, Kenneth B. (1996) *The Making of Modern Japan*, 2nd ed. Lexington, MA: D. C. Heath.

Ramm, Agatha (1967) *Germany, 1789–1919: A Political History.* London: Methuen.

RETALLACK, JAMES (1992) "Wilhelmine Germany." In *Modern Germany Reconsidered, 1870–1945*, edited by Gordon Martel. London: Routledge.

RICH, NORMAN (1992) *Great Power Diplomacy, 1814–1914*. New York: McGraw-Hill.

ROBBINS, KEITH (1994) *The Eclipse of a Great Power: Modern Britain, 1870–1992*, 2nd ed. London: Longman.

ROGGER, HANS (1983) *Russia in the Age of Modernisation and Revolution, 1881–1917*. New York: Longman.

RÖHL, JOHN C. G. (1995) "Germany." In *Decisions for War, 1914*, edited by Keith Wilson. New York: St. Martin's Press.

ROSECRANCE, RICHARD (1995) Overextension, Vulnerability, and Conflict: The "Goldilocks Problem" in International Strategy. *International Security* **19**:145–163.

ROSECRANCE, RICHARD, AND ARTHUR A. STEIN (1993) *The Domestic Bases of Grand Strategy*. Ithaca, NY: Cornell University Press.

ROSECRANCE, RICHARD, AND ZARA STEINER (1993) "British Grand Strategy and the Origins of World War II." In *The Domestic Bases of Grand Strategy*, edited by Richard Rosecrance and Arthur A. Stein. Ithaca, NY: Cornell University Press.

RUBINSTEIN, W. D. (1998) *Britain's Century: A Political and Social History, 1815–1905*. London: Arnold.

SAGAN, SCOTT D. (1988) "The Origins of the Pacific War." In *The Origin and Prevention of Major Wars*, edited by Robert I. Rotberg and Theodore J. Rabb. Cambridge: Cambridge University Press.

SCHMIDT, GUSTAV (1990) "Contradictory Postures and Conflicting Objectives: The July Crisis." In *Escape into War? The Foreign Policy of Imperial Germany*, edited by Gregor Schöllgen. Oxford: Berg.

SCHULZINGER, ROBERT D. (1998) *U.S. Diplomacy Since 1900*, 4th ed. Oxford: Oxford University Press.

SCHWELLER, RANDALL L. (1994) Bandwagoning for Profit: Bringing the Revisionist State Back In. *International Security* **19**:72–107.

SETON-WATSON, CHRISTOPHER (1967) *Italy from Liberalism to Fascism: 1870–1925*. London: Butler & Tanner.

SETON-WATSON, HUGH (1967) *The Russian Empire, 1801–1917*. New York: Oxford University Press.

SINGER, J. DAVID (1961) "The Level-of-Analysis Problem in International Relations." In *International Politics and Foreign Policy: A Reader in Research and Theory*, edited by James N. Rosenau. New York: Free Press.

SKED, ALAN (1989) *The Decline and Fall of the Habsburg Empire, 1815–1918.* New York: Longman.

SMALL, MELVIN (1996) *Democracy and Diplomacy: The Impact of Domestic Politics on U.S. Foreign Policy, 1789–1994.* Baltimore, MD: Johns Hopkins University Press.

SNYDER, GLENN H., AND PAUL DIESING (1977) *Conflict Among Nations: Bargaining, Decision Making, and System Structure in International Crises.* Princeton, NJ: Princeton University Press.

SNYDER, JACK (1991) *Myths of Empire: Domestic Politics and International Ambition.* Ithaca, NY: Cornell University Press.

SNYDER, RICHARD C., H. W. BRUCK, AND BURTON SAPIN (1954) Decision-Making as an Approach to the Study of International Politics. Monograph no. 3, Foreign Policy Analysis Project Series. Princeton, NJ: Princeton University.

SPRING, D. W. (1988) "Russia and the Coming of War." In *The Coming of the First World War*, edited by R. J. W. Evans and Hartmut Pogge von Strandmann. Oxford: Oxford University Press.

STEIN, ARTHUR A. (1993) "Domestic Constraints, Extended Deterrence, and the Incoherence of Grand Strategy: The United States, 1938–1950." In *The Domestic Bases of Grand Strategy*, edited by Richard Rosecrance and Arthur A. Stein. Ithaca, NY: Cornell University Press.

STEINBRUNER, JOHN D. (1974) *The Cybernetic Theory of Decision: New Dimensions of Political Analysis.* Princeton, NJ: Princeton University Press.

STEINER, ZARA S. (1977) *Britain and the Origins of the First World War.* New York: St. Martin's Press.

STERN, FRITZ (1967) "Bethmann-Hollweg and the War: The Limits of Responsibility." In *The Responsibility of Power*, edited by Leonard Krieger and Fritz Stern. Garden City, NY: Doubleday.

STONE, NORMAN (1966) Hungary and the Crisis of July 1914. *Journal of Contemporary History* **1**:147–164.

TAUBMAN, WILLIAM (1982) *Stalin's American Policy: From Entente to Détente to Cold War.* New York: Norton.

THAYER, JOHN A. (1964) *Italy and the Great War: Politics and Culture, 1870–1915.* Madison: University of Wisconsin Press.

TOMBS, ROBERT (1996) *France, 1814–1914.* New York: Longman.

TROUT, B. THOMAS (1975) Rhetoric Revisited: Political Legitimation and the Cold War. *International Studies Quarterly* **19**:251–284.

VASQUEZ, JOHN (1993) *The War Puzzle*. Cambridge: Cambridge University Press.

VEITH, JANE KAROLINE (1996) "The Diplomacy of Depression." In *Modern American Diplomacy*, edited by John M. Carroll and George C. Herring. Wilmington, DE: Scholarly Resources.

WALTZ, KENNETH N. (1979) *Theory of International Relations*. Reading, MA: Addison-Wesley.

WATT, DONALD CAMERON (1989) *How War Came: The Immediate Origins of the Second World War, 1938–1939*. New York: Pantheon.

WEHLER, HANS-ULRICH (1985) *The German Empire, 1871–1918*. New York: Berg.

WESTWOOD, J. N. (1993) *Endurance and Endeavour: Russian History, 1812–1992*. New York: Oxford University Press.

WILLIAMS, GLYN, AND JOHN RAMSDEN (1990) *Ruling Britannia: A Political History of Britain, 1688–1988*. New York: Longman.

WILLIAMSON, SAMUEL R., JR. (1969) *The Politics of Grand Strategy: Britain and France Prepare for War, 1904–1914*. London: Ashfield Press.

WILLIAMSON, SAMUEL R., JR. (1983) "Vienna and July 1914: The Origins of the Great War Once More." In *Essays on World War I: Origins and Prisoners of War*, edited by Samuel R. Williamson, Jr., and Peter Pastor. New York: Columbia University Press.

WILLIAMSON, SAMUEL R., JR. (1988) "The Origins of World War I." In *The Origins and Prevention of Modern Wars*, edited by Robert I. Rotberg and Theodore K. Rabb. New York: Cambridge University Press.

WILLIAMSON, SAMUEL R., JR. (1991) *Austria-Hungary and the Origins of the First World War*. London: Macmillian Press.

WILSON, KEITH (1995) "Britain." In *Decisions for War, 1914*, edited by Keith Wilson. New York: St. Martin's Press.

WRIGHT, GORDON (1995) *France in Modern Times*, 5th ed. New York: Norton.

YERGIN, DANIEL (1977) *Shattered Peace: The Origins of the Cold War and the National Security State*. Boston: Houghton Mifflin.

YOUNG, JOHN W. (1997) *Britain and the World in the Twentieth Century*. London: Arnold.

YOUNG, ROBERT J. (1996) *France and the Origins of the Second World War*. New York: St. Martin's Press.

ZELDIN, THEODORE (1973) *France, 1848–1945, Volume 1: Ambition, Love, and Politics*. Oxford: Clarendon Press.

# How Decision Units Shape Foreign Policy:

## A Theoretical Framework

*Margaret G. Hermann*

*Maxwell School, Syracuse University*

Two questions must be addressed if we are going to get inside the "black box" of government to understand the relevance of leadership to foreign policymaking: (1) What types of actors make foreign policy decisions? (2) What is the effect of these decision units on the resulting foreign policy? An examination of how governments and ruling parties around the world make foreign policy decisions suggests that authority is exercised by an extensive array of different entities. Among those making policy are prime ministers, presidents, party secretaries, standing committees, military juntas, cabinets, bureaucracies, interagency groups, legislatures, and loosely structured revolutionary coalitions. When we contemplate engaging in systematic comparisons of governmental decision-making bodies across and within countries, the number of possibilities becomes formidable.

The premise of this special issue is that there is a way of classifying these decision units that can enhance our ability to account for governments' behavior in the foreign policy arena. In particular, three types of decision units are found in the various political entities listed above: the powerful leader, the single group, and the coalition of autonomous actors. The decision units framework presented here is intended to assist the researcher in ascertaining when each of these types of units is likely to be involved in making a foreign policy decision as well as how the structure and process in the unit can affect the nature of that decision.

Although we recognize there are numerous domestic and international factors that can and do influence foreign policy behavior, these influences are necessarily channeled through the political apparatus of a government that identifies, decides, and implements foreign policy. Policy is made by people configured in various ways depending on the nature of the problem and the structure of the government. Indeed, we argue that there is within any government an individual or a set of individuals with the ability to commit the resources of the society and, when faced with a problem, the authority to make a decision that cannot be readily reversed. We call this set of decision makers the "authoritative decision unit" and seek to understand how it shapes foreign policy decision making across diverse situations and issues as well as different political settings.

We are interested in one stage of the foreign policymaking process: the point at which members of the authoritative decision unit select a particular course of action, that is, make a choice. Even though we are aware that the actual process of choice may not be a clear occurrence, that key decisions and those who make them are constrained by available inputs, and that subsequent implementation of a decision may lead to distortion, knowledge about how decisions are made is a powerful source of insight into what complex entities, such as governments, do. By learning about how foreign policy decisions are made, we gain information about the intentions and strategies of governments and how definitions of the situation are translated into action.

## CORE ASSUMPTIONS OF THE DECISION UNITS APPROACH

The decision units approach described here builds upon a growing body of research on foreign policy decision making (for reviews of this literature see, e.g., 't Hart, 1990; Maoz, 1990; Vertzberger, 1990; Bender and Hammond, 1992; Khong, 1992; Welch, 1992; Caldwell and McKeown, 1993; Evans, Jacobson, and Putnam, 1993; Hagan, 1994; Kupchan, 1994; Hermann and Kegley, 1995; Hudson, 1995; 't Hart, Stern, and Sundelius, 1997; George and George, 1998; Stern and Verbeek, 1998; Sylvan and Voss, 1998; Allison and Zelikow, 1999; Rosati, 2000). These works overview decision-making "models" that focus on bureaucratic politics, group dynamics, presidential advisory systems, governmental politics, leadership, coalition politics, and the strategies for dealing with domestic opposition. The decision units framework attempts to integrate this extant research literature.

The approach is grounded in three assumptions about foreign policymaking that merit some discussion. (1) These so-called models of decision making examine decision units that are found in most governments, yet researchers have wanted to declare one a winner—"the" explanation for how foreign policy decisions are made. The literature does not facilitate our understanding of foreign policymaking by treating them as separate, complementary frameworks for explaining the essence of decision. (2) Much of the decision-making literature, as well as that in international relations, has focused on the constraints that limit what decision units can do, failing to take into account the variety of ways in which those involved in policymaking can shape what happens. Decision units are often active participants in the making of foreign policy. (3) We are intent on developing a framework that facilitates scholars exploring how decisions are made in all types of countries. To date models of foreign policy decision making have had a distinctly U.S. flavor. As a result, the models have not fared as well when extended to non-U.S. settings, particularly to nondemocratic, transitional, and less developed polities (see, e.g., Korany and Dessouki, 1991; Hagan, 1993; Stern and Verbeek, 1998). Indeed, "the U.S. bias" in the decision-making literature has made it difficult to generalize to other countries and has given researchers blind spots regarding how decisions are made in governments and cultures not like the American. Before explicating our approach further, let us examine in more detail the reasons for our first two assumptions, in turn, noting how our desire to be comparative has shaped the more integrated approach advocated here.

## *Viewing the Models as Contingent*

Since Allison's (1971) seminal work exploring which of three models of decision making was most useful in understanding the choices American policymakers made in the Cuban missile crisis, scholars engaged in foreign policy analysis have tended to view the models—and, in turn, the types of decision units—in competition with one another as explanations of governments' actions in international relations. If the goal is the development of a comparative framework for understanding foreign policy decision making, we wonder if this strategy is the most appropriate. We do not dispute that there are alternative decision units and processes. At issue is the conclusion that any one decision unit is generally more valid than the others in explaining foreign policy decision making. In our view, all the various decision-making models in the literature have merit. Our strategy is to identify the theoretical conditions under which each set of decision dynamics is more likely to occur. For each type of decision unit, we want to specify the variables that lead to one particular process as opposed to the others. For example, what factors predispose a cabinet to engage in bureaucratic politics as opposed to groupthink? Thus, instead of exploring which set of variables is most potent in explaining a particular foreign policy action or applying "alternative cuts" to a case assuming that one will have more explanatory power than the others, our approach posits that it is theoretically possible to determine the conditions under which each of the models is most applicable and assumes that all models will apply in certain, specifiable situations. In other words, the various decision-making models are all relevant to understanding the foreign policymaking process; the decision units framework suggests in what political structures, kinds of problems, and situations each type is expected to prevail.

This kind of logic is especially important if decision theories are to have cross-national validity. As Hagan noted earlier in this special issue, much conventional international relations research presumes that decision-making processes are determined by basic national and political system characteristics. Thus, Western democracies are viewed as having pluralistic processes while authoritarian political systems are seen as hierarchical and highly cohesive, and the policies in Third World polities are determined by the predominant leader's personal predispositions. In contrast, scholars with area expertise have shown the weaknesses in this argument. For example, states with predominant leaders have at times been governed more by coalitions of interests and group dynamics than by the views and goals of a single actor, while highly bureaucratized governments have seen a dominant leader centralize authority and push a particular ideology or cause (see, e.g., Weinstein, 1972; Lincoln and Ferris, 1984; Vertzberger, 1984; Korany and Dessouki, 1991; Snyder, 1991; Hermann and Kegley, 1995). These latter insights have helped to guide our development of a contingency model of foreign policy decision making. They caution against assuming that certain decision-making processes are a direct function of basic national attributes or the structure of the political system. Furthermore, they suggest that the nature of the decision unit is just as likely to vary within a single country as between different types of nations.

## *Considering the Full Range of Decision Processes*

Our second assumption builds from our observation that there are a variety of potential outcomes that can result from the decision process. What happens within a decision unit in the decision-making process can lead to an array of different kinds of outcomes, indicating a need to move beyond characterizing the outcomes of decisions as simply "political resultants." Consider that in some cases there is a decision not to act or an inability to mount a new policy initiative while in other cases the decision dynamics may propel one party's position to dominate, leading to more extreme action than most would have desired. Somewhere between these two outcomes of deadlock and strong forceful action are more complex situations where policies are "watered down" as a result of internal bargaining and compromise or one party moderates its position in order to let another "save face."

Our point here is that a comparative explanation of foreign policy needs to recognize that decision-making dynamics do not have a direct, singular impact on foreign policy. Rather, they can produce various results from consensus to deadlock, from compromise to domination by one individual or faction. Our explanations need to account for both the "push" and "pull" of these decision-making dynamics—for when they are likely to moderate or diminish the nature of a proposed response as well as when they will exacerbate the situation and produce a stronger action than might otherwise have been chosen. Not only will clearer conceptual efforts enable us to better judge the effects of decision-making processes, but they will make it possible to elaborate the linkages between decision units and decision outcomes.

To date, when the foreign policymaking literature has considered the outcomes of the decision process, the tendency has been to emphasize *pathological* decision-making behaviors (e.g., Allison, 1971; Halperin, 1974; George, 1980; Janis, 1982; C. Hermann, 1993; Stern and Verbeek, 1998). In other words, the emphasis has been on explicating decision-making models that account for suboptimal outcomes reflective of decision makers' failure to respond effectively to international pressures and domestic problems. Interest has centered around considering decision-making processes only when they seem to force dramatic deviations from the presumed norm of rational policymaking that involves being open to and understanding international and domestic constraints.

Regardless of whether actual decision-making processes conform to the ideals of rationality (about which there is debate), it does not follow that one must argue that these processes are always pathological or that they invariably lead to suboptimal decision outcomes. Indeed, we need models of decision making that not only reveal how things can go wrong, but reliably explain the process operating in a range of decisions irrespective of the evaluative assessment of those procedures and the results they produce. In particular, we want to understand the contingencies that increase the likelihood that a decision unit (a) adequately recognizes stimuli from its environment and (b) achieves timely collaboration among its members so that they can reach an agreement and engage in meaningful action.

Once again, such elaboration has particular importance for cross-national foreign policy research efforts. The premature connection of particular types of decision units and processes with specific kinds of countries or political systems can lead to distortions in our explanations of their foreign policies. We want to avoid the presumption that certain defects (or virtues) are inherent in particular political structures or philosophies—for example, that democratic decision making is *always* more reactive and incoherent than decision making in authoritarian regimes, or that the actions of rogue states are reckless and out of touch with any kind of reality. An understanding of the conditions conducive to particular kinds of processes and outcomes would not only improve our understanding of how far countries' foreign policy is likely to stray from the optimal, but also presumably help scholars avoid the application of simplistic stereotypes regarding what those states are likely to do.

## A Decision Units Approach to Foreign Policy Decision Making

Building on the previous discussion, our proposed framework has several components: (1) it views decision making as involving responding to foreign policy problems and occasions for decision; (2) it focuses on three types of authoritative decision units; (3) it defines the key factors that set into motion alternative decision processes; and (4) it links these alternative decision processes to particular outcomes. When combined, these components articulate a contingency approach to the study of foreign policy decision making. Figure 1 diagrams the interrelation of the various components. Although space does not permit explaining each component in detail, we will provide an overview of the framework here. The theories and decision logics embedded in each of the different types of decision units will be explicated further in the next three pieces in this special issue. (For more detail on the development of the ideas presented here see Hermann and Hermann, 1982; Hermann and Hermann, 1985; Hermann, Hermann, and Hagan, 1987; Hermann and Hermann, 1989; Stewart, Hermann, and Hermann, 1989; Hagan, 1993; C. Hermann, 1993; M. Hermann, 1993; Hermann and Hagan, 1998.)

## *Inputs to the Decision Units Framework*

What triggers governments to make foreign policy decisions that, in turn, prod powerful leaders, single groups, and coalitions into action? What is it about the political setting that leads one or the other of these different types of decision units to assume authority for making a decision at any point in time? How do we know which of the three types of decision units should be the focus of our attention in studying a particular event? The answers to these questions form the inputs for the application of the decision units approach. They start the framework in motion. Of interest is what precipitates a foreign policy decision and a particular decision unit taking action. The inputs to the framework represent the stimuli from the international and domestic environments to which the authoritative decision unit is responding.

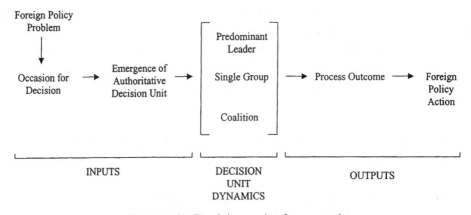

FIGURE 1. Decision units framework

***Problems trigger decisions.*** Discussions with policymakers and policy analysts suggest that they respond to problems embedded in situations (see, e.g., George, 1993; Galvin, 1994; Kruzel, 1994). Policymakers have goals and objectives they believe are important and want to achieve during their administrations; agendas for foreign policy are formed around these plans. But often as they begin to take action on such goals and objectives, they encounter problems in their domestic and international environments that challenge what they want to do. Their agendas can also be changed as they are forced by situations happening elsewhere in the world to attend to issues not necessarily among their priorities. As a result, governments take action when policymakers perceive a problem in foreign policy that they believe they can or need to influence. Decisions are sought to deal with problems.

By problem is meant a perceived discrepancy between present conditions and what is desired. As this definition implies, problems are subjective. The nature of the problem—and, indeed, whether one exists at all—depends on policymakers' perceptions. Accordingly, different policymakers, and different governments, may observe the same state of affairs but recognize distinctive problems or no problem at all. Moreover, problems can pose opportunities as well as difficulties for policymakers and governments. A perceived discrepancy that suggests action can lead to a more preferable condition offers policymakers an opportunity; a perceived discrepancy that denotes things are changing or could change for the worse challenges what policymakers are doing and can become threatening.

A problem is recognized when policymakers state that something is wrong, needs attention, or presents an opportunity for gain if action is taken. Thus, expressions by representatives of a regime or administration about some current difficulty or potential opportunity indicate the recognition of a problem. Governments often organize their foreign policy bureaucracies to allow policymakers to monitor various regions of the world and certain issue areas for problems. For example, a South Asian Desk includes people whose job is to attend to what is happening in countries in that part of the world and to note when events or these governments' actions pose a difficulty or opportunity for achieving certain goals. A Bureau for Inspection and Verification oversees whether there are treaty violations by other countries with regard to certain weapons' systems that could threaten their government's programs and initiatives. We propose that when problems are recognized, decision units are generally convened to deal with them.

Thus, in exploring how foreign policy decisions are made, we start with a problem that needs addressing. Problems are the trigger or reason for engaging the decision units framework. Not only is the foreign policy problem the initial stimulus or input into the framework, attributes of problems provide us with helpful information in identifying the authoritative decision unit and some ideas about the options under review. Although we know that considerable interest exists in understanding how societies and governments decide what problems to address and what priority to give them on their collective agenda, such is not the focus of attention of the decision units approach. We are studying who deals with problems once identified and how the process they use affects the nature of the decision. When policymakers have recognized a foreign policy problem, we want to determine who will be able to commit the resources of the government and how that individual or those entities go about making a decision.

***Occasions for decision.*** Foreign policy problems arise episodically and often lead to a series of decisions. Policymakers generally do not deal with a problem by making a single decision and then sit back to await a response. Problems tend to get structured into a string of decisions that involve different parts of the government's foreign policy machinery. Consider as an illustration the British response to the Argentinean invasion of the Falkland Islands. The response consisted of a series of decisions made in the British cabinet, defense ministry, parliament, and foreign ministry. Different aspects of the problem were dealt with by policymakers in these various institutions—general guidelines for policy were developed by the cabinet, troop movements were defined by the defense ministry, cabinet policy was ratified by the Parliament, and diplomatic moves in the United Nations and elsewhere were determined by the foreign ministry (Franks, 1983; Hastings and Jenkins, 1983; Lebow, 1985). In effect, in responding to a foreign policy problem governments often are involved in a sequence of decisions.

Each time policymakers formulate a question about a recognized foreign policy problem that needs answering and arrange for someone or somebody to respond to it, we have an occasion for decision. Occasions for decision represent the instances in coping with a problem when the policymakers are faced with making a choice. They are those points in the decision process when there is a felt need by those involved to take action even if the action is the choice to do nothing or to search for more information. As a result, problems often include a number of occasions for decision that may be addressed across time by the same decision unit or by all three types of decision units.

Occasions for decision are usually perceived by policymakers as questions that need to be addressed. The questions that drive occasions for decision generally take one of three forms: (1) queries about whether action is needed in relation to this problem (e.g., are Iraq's troop movements toward the Kuwaiti border something we need to take action on at this time?); (2) queries seeking possible solutions to a problem (e.g., what should we do about the Iraqi troop movements toward the Kuwaiti border?); or (3) queries about whether one or more proposals for dealing with this aspect of the problem should be adopted (e.g., should we send troops to counter the Iraqis or should we go to the United

Nations Security Council with a resolution condemning their movement?). The first two types of occasions for decision lead to policy declarations in which broad policy directions are stipulated and goals and objectives are set. The third type of occasion for decision leads to a strategic decision in which a particular action is chosen and resources are committed.

Occasions for decision that call for policy declarations and strategic actions are distinguished in the present research from other types of occasions for decision. With the decision units framework we are interested in examining occasions for decision that lead to authoritative actions on the part of the government in dealing with a perceived foreign policy problem. We want to understand the processes that affect the commitment of a government's resources and its choice of policy. Of less interest are those occasions for decision that are focused on searching for more information about a problem, implementing previous authoritative decisions, or the ratification of a decision. Although we recognize that these latter types of occasion for decision have implications for policy declarations and strategic choices by providing more differentiated input into the selection process, by shaping the consequences or reactions to a policy choice, or by distorting how any decision is carried out, they are only the focus of attention in the decision units framework if they have resulted in an occasion for decision that calls for an authoritative decision on what the government is going to do or not do with regard to the problem at hand.

Typically an occasion for decision that calls for a policy choice or authoritative decision can be detected in reports that policymakers are looking for a means of handling a problem or considering whether or not to act on a problem they perceive in the international arena. Such occasions for decision are also recognizable in reports that policymakers are discussing a particular option or that there is a debate among policymakers about one or more options for dealing with a problem. Moreover, indications that there are disagreements among policymakers on how a problem should be defined or about what alternatives are feasible are suggestive that an authoritative decision may be in process or required soon.

Whenever the question that forms the basis for an occasion for decision changes, we have a new occasion for decision and the possibility for a new decision unit to address it. The question can shift for a variety of reasons. For example, the foreign recipient of a government's action can fail to respond in the expected manner; new information can cause a reinterpretation of the problem; issues can be raised in implementing a decision that bring the former decision into question; one of the major participants in the decision unit can change his, her, or its preferences; efforts can be made to overcome a deadlock by changing participants in the decision unit; a minority who lost out in influencing the decision earlier can resurface the issue again. Hence, the occasion for decision performs a similar function in the decision units framework as a single frame of film does in a motion picture. The single frame is the building block from which a continuous strip of film is made. By focusing on occasions for decision, we take snapshots of the decision-making process at various points in time as policymakers attempt to deal with a problem. Together, just like frames in a movie, the occasions for decision form an episode such as the Cuban missile crisis, the Gulf War, or the Mexican peso crisis.

The occasion for decision provides observers and analysts a basic unit of analysis for studying how policymakers and governments deal with foreign policy problems. It facilitates isolating and examining the sequence of decisions that are made in handling such problems by breaking the sequence into its parts. In rebuilding the sequence we learn about the flow of decisions and who was involved in which decisions with what consequences to the decision process and actual choice. In effect, *each time* there is a new occasion for decision as policymakers cope with a foreign policy problem, we reapply the decision units framework to see if the decision unit has changed or some aspect of the setting within the decision unit has changed.

***The authoritative decision unit.*** At the apex of foreign policy decision making in all governments or ruling parties is a group of actors—the authoritative decision unit—who, if they agree, have both the ability to commit the resources of the government in foreign affairs and the power to prevent other entities within the government from overtly reversing their position. The unit having this authority in a country may (and frequently does) vary with the nature of the problem. For issues of vital importance to a country, the highest political authorities often constitute the decision unit; there is a contraction of authority to those most accountable for what happens. For less dramatic, more technical issues, the ultimate decision unit generally varies depending on the type of problem the government is facing (military, economic, diplomatic, environmental, scientific, and so on). In governments where policy normally involves multiple bureaucratic organizations, the problem may be passed among different units—within one agency, across agencies, or between interagency groups. The basic point here is that for most foreign policy problems and occasions for decision, some person or collection of persons come together to authorize a decision and constitute for that issue at that point in time the authoritative decision unit.

As we observed earlier, an examination of the various decision-making models that have been proposed in the literature indicates that there are, in essence, three types of possible authoritative decision units. They are:

1. PREDOMINANT LEADER: A single individual who has the ability to stifle all opposition and dissent as well as the power to make a decision alone, if necessary.

2. SINGLE GROUP: A set of individuals, all of whom are members of a single body, who collectively select a course of action in consultation with each other.
3. COALITION OF AUTONOMOUS ACTORS: The necessary actors are separate individuals, groups, or representatives of institutions which, if some or all concur, can act for the government, but no one of which by itself has the ability to decide and force compliance on the others; moreover, no overarching authoritative body exists in which all these actors are members.

This categorization is considered both mutually exclusive and exhaustive. The actors who make authoritative decisions for governments in the foreign policy arena should correspond to one of these three configurations. Consider some examples of potential authoritative decision units from the early twenty-first-century international scene that represent each of these types: Cuba's Fidel Castro and Iraq's Saddam Hussein are predominant leaders; single groups include the British cabinet and the Standing Committee of the Chinese Communist Party; and coalitions of autonomous actors are in evidence in Iran where foreign policy can only result from the interaction of the more moderate forces led by President Khatami and the more conservative forces led by Ayatollah Khamenei as well as in Indonesia where the president and vice-president are drawn from different parties and points of view.

## *Determining the Authoritative Decision Unit for an Occasion for Decision*

Figure 2 shows the sets of factors we consider in determining which of the three types of decision units—predominant leader, single group, or coalition—will have ultimate authority to respond to a particular occasion for decision. (See Hermann and Hermann, 1989, for a more detailed figure and discussion of what constitutes each of the factors.) Before we can begin to explore the ways in which decision units can affect policy, we need to ascertain the type of decision unit that has the authority to commit the resources of the government for a specific occasion for decision. Thus, once we have decided that policymakers recognize a foreign policy problem and are faced with an occasion for decision that calls for a declaration of policy or strategic choice, we have to determine who will make the decision.

The factors in Figure 2 take into account both formal and informal structures of government. At issue is where in the government is the problem under discussion and this specific occasion for decision likely to receive attention. To answer this question we need to determine how the government is structured by law as well as consider the norms that have arisen around these institutional arrangements. The questions in the figure focus first on the formal structures of

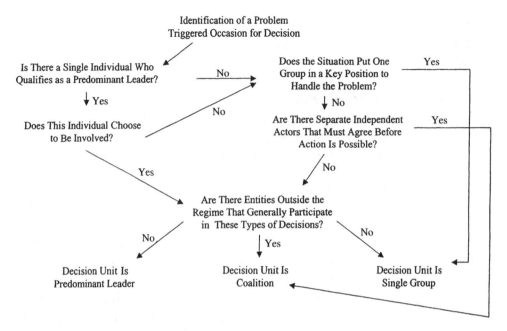

FIGURE 2. Factors involved in determining the nature of the authoritative decision unit for an occasion for decision

governance and then on the informal structures that may be in effect for a particular occasion for decision. We are interested in whether there is in the regime's leadership one individual with the power and authority to commit or withhold the resources of the government with regard to the problem at hand (a predominant leader), a single group that is responsible for dealing with problems like the current one, or two or more separate actors (individuals, groups, organizations) that must agree before the resources of the government can be committed for such problems (a coalition).

***Conditions favoring a predominant leader.*** The decision unit for any occasion for decision is likely to be a predominant leader if the regime has one individual in its leadership who is vested with the authority—either by the constitution, law, or general practice—to commit or withhold the resources of the government with regard to foreign policy problems. A monarchy is an illustration of this kind of predominant leader as is a presidential political system in which the president is given authority over foreign policy.

The decision unit can also be a predominant leader if the foreign policy machinery of the government is organized hierarchically with one person located at the top of the hierarchy who is ultimately accountable for any decisions that are made. As Harry Truman said about the American presidency, "The buck stops here." Moreover, if a single individual has control over the various forms of coercion available in the society and, as a result, wields power over others, the decision unit can be a predominant leader. Dictatorships and authoritarian regimes fall into this category and often have predominant leaders dealing with foreign policy matters.

If we ascertain there is a predominant leader at this point in time, we need to determine whether or not he or she chooses to exercise that authority. We know, for instance, that even though Franco was a predominant leader in Spain during the 1950s and 1960s, he turned over much of his foreign policymaking authority to his foreign minister (Gunther, 1988). How can we assess if the single powerful leader is dealing with a specific occasion for decision? The literature on political leadership and foreign policy decision making suggests there are at least six conditions when such leaders are likely to exercise their powers (see Hermann, 1976, 1988; Holsti, 1976; Greenstein, 1987). These six conditions include certain types of situations that guarantee involvement—high-level diplomacy, crisis events—and particular aspects of the leaders' personalities that push them to want control over what happens—interest, expertise, and techniques for managing information and resolving disagreements.

If an occasion for decision involves high-level diplomacy where the predominant leader is expected to be a part of what happens, chances increase that he or she will take charge and exercise authority. Summit meetings are a case in point. Of necessity the predominant leader will be present and participating actively in the policymaking process. When faced with a crisis situation, predominant leaders are also likely to become involved in what is happening in foreign policy even if they are not generally involved. Literature on organizations and bureaucracies (e.g., Hermann, 1972; Lebow, 1981; Hampson, 1988; 't Hart, 1990) indicates that there is a contraction of authority during situations that are critical to the survival of the government. Policymakers at the top take part in the decision-making unit because their positions and policies often are under challenge. Their interest in overseeing the process heightens as their accountability for what happens increases.

Predominant leaders can be more and less prone themselves to want to take charge. Studies of political leadership (e.g., Barber, 1977; Burns, 1978; Hermann, 1984, 1988; Preston, 2001) have shown that predominant leaders are more likely to become involved in what is happening the greater their general interest in foreign policy and the more extensive their experience or expertise. Such leaders tend to gravitate toward the area of policy where they feel comfortable. These leaders are likely to choose to follow the issues arising in foreign policy and to help define the agenda and problems that are relevant to their administration. It is almost second nature to work on foreign policy problems first. Similarly, although a predominant leader may not generally be interested in foreign policy or have a lot of foreign policy experience, he or she may be particularly interested, or have expertise, in a specific type of foreign policy problem and insist on being involved when those interests or problems are considered. For example, Dwight Eisenhower's special interest in controlling the spread of nuclear weapons through developing peaceful uses for atomic energy led him to want to take charge of any decisions the American government made regarding this topic.

Predominant leaders' preferred ways of dealing with advisers can also suggest whether or not they will generally exercise authority (see, e.g., Burke and Greenstein, 1991; Hermann and Preston, 1994; George and George, 1998). If such leaders organize those around them into a team and are interested in serving as the center of the team, they are likely to be involved in most foreign policy occasions for decision that emanate from the government. These predominant leaders maintain control by becoming the hub of the information network, having a better overall picture of what is going on than any one player on the team. Contrast this scenario with that for predominant leaders who prefer to receive information after it has been digested and sifted by advisers and those in the bureaucracy. These predominant leaders only want to deal with the "most important" decisions or those that make it through the hierarchy. Moreover, such leaders often make foreign policy by issuing directives for others to interpret. For predominant leaders with this latter leadership style, some occasions for decision will be dealt with by other policymakers. This type of predominant leader will more often be the authoritative decision unit for occasions for decision that focus on goal-setting or defining objectives than for those that involve deciding on particular strategies.

***Conditions favoring a single group.*** If the government is not structured around a single individual, there may be a designated group that is responsible for dealing with the occasion for decision under consideration. Such a key group can take one of several forms depending on its location in the government and the nature of the problem stimulating the occasion for decision.

There may be one particular group whose role is to deal with the type of occasion for decision that is facing the government. If so, the authoritative decision unit becomes a single group rather than a predominant leader. The Politburo in the former Soviet Union and the Joint Chiefs of Staff in the American government are illustrations of such single groups. Although responsible for foreign policy problems in their respective governments, these two groups differ in the extent of their authority. The Politburo appears to have been in on most occasions for decision that involved foreign policy; the Joint Chiefs of Staff focus only on military problems and, in turn, strategic choices rather than policy declarations. Often such groups develop around recurrent problems that the members all have a stake in solving and have some expertise in resolving. It is important in considering whether or not there is such a group to determine if certain people are brought together in a face-to-face setting in most instances when problems of this nature arise.

Governments also have in place groups that have as their mission handling crisis problems when they arise. Such groups become key to decision making when the problem is critical to the government. Thus, defense and foreign ministries have groups whose purpose is to respond to crises. Situation or war rooms are manifestations of such organizational structures. Personnel are trained to monitor and to consider devising standard operating procedures for coping with such problems. Moreover, the literature on crisis management (e.g., Lebow, 1981; Kleiboer, 1998; Stern, 1999) suggests that governments may create groups with the sole function of considering ways to defuse a particular situation. The American ExCom in the Cuban missile crisis is an example of this type of group. Its members, who were for the most part also cabinet members, were brought together to deliberate on the options the U.S. government had to deal with the so-called offensive Soviet missiles in Cuba. They no longer functioned as a group when the crisis was over.

Furthermore, for the current occasion for decision there may be a key group in a relevant bureaucracy that is charged with handling such problems. Bureaucracies are often organized around problem areas or regions of the world. For example, foreign ministries could have a Middle East Bureau that focuses on problems in that part of the world or they could have an agency paying attention to human rights issues. At times, there are also standing interagency committees that are charged with developing policies regarding security or economic problems that coordinate among bureaucratic organizations. Remember that to be considered a single group there need to be two or more people who interact directly with each other and collectively reach a decision. All persons necessary to committing the resources of the government with regard to the particular occasion for decision must be members of the group for us to have a single group as the authoritative decision unit. In effect, in the single group decision unit there is a collective, interactive (generally face-to-face) decision process in which all members participate.

*Conditions favoring coalitions.* At times in governments faced with an occasion for decision, the authoritative decision unit is composed of multiple autonomous actors. That is, two or more entities (e.g., individual leaders, groups of policymakers, bureaucratic agencies, interest groups) have the power to commit or withhold the resources of the government and none can allocate such resources without the concurrence of the other(s). Thus, in locating what is the authoritative decision unit for a particular occasion for decision, we need to consider whether or not there are separate, independent actors who must work together in making a decision or nothing will happen. As an illustration, let us examine the Iranian government during the period when the students took over the American embassy and held the personnel there hostage. In order for the Iranian government to make a decision regarding release or transfer of the hostages, the following entities had to concur: Ayatollah Khomeini, the mullahs in the Revolutionary Council, the moderates in the government ministries, and the students who took the embassy. None could commit the resources of the government without the consent of the others. This type of decision unit consists of multiple independent actors who form into a single coalition or multiple coalitions in arriving at a decision.

The entities comprising the multiple independent actors can be from outside the government as well as from within the government. We have already suggested one instance of this with the Iranian students who took the American hostages. Several other examples may help make the point. Consider the leadership of the Catholic Church in a strongly Catholic country that may need to be consulted by the government in advance over policy issues central to its doctrine, or a politically active military that may be involved in decision making in a polity marked by recurrent military intervention. To be included as part of this type of decision unit, the actor outside the government must participate in the decision making on the issue not just in the implementation of the decision. Such actors need to have the possibility of withholding the resources necessary for action *if not consulted* as part of the decision-making process. These nongovernmental actors must regularly engage in the decision process on certain issues or have the power to veto or reverse government decisions to be considered part of the decision unit. Their veto or override authority may be based in law but more probably is the result of the entity's ability to block or alter implementation of a decision because it controls key resources or has moral authority in particular matters.

Multiple autonomous actors can also include foreign governments or their representatives, multinational corporations, or other international organizations (e.g., the World Bank or United Nations). The most common occurrence of this type of actor is in a government dominated militarily or economically by another country. Eastern European countries before the fall of the Soviet Union often had to consider the interests of the government of the Soviet Union in making their foreign policy. It is well known that Ferdinand Marcos of the Philippines regularly consulted U.S. policymakers on issues of foreign policy that involved the United States in order to ensure future economic aid. To be considered part of the decision unit, such actors must routinely exercise control of the government's decisions regarding the particular resources involved in the occasion for decision.

Even if the formal structures of the government suggest that the authoritative decision unit for a particular occasion for decision is a predominant leader or single group, the presence of relevant actors outside the government can change the nature of the decision unit. As the example with Marcos implies, a predominant leader sought out the participation of others in certain decisions. Fearing an overthrow, a cabinet (single group) may consult the military regarding decisions in which it might have an interest in order to prevent such from happening. The dispersion of power and influence in a government and society may change the authoritative decision unit from what appears formally the case to an informal set of actors who must collaborate for anything meaningful to be decided.

***Some boundary issues in determining the authoritative decision unit.*** Experience in determining the nature of the authoritative decision unit for occasions for decision has posed some classification issues that fall at the boundaries of the definitions of the particular units. For example, is a leader-advisory group an instance of a predominant leader or single-group decision unit? Is a coalition cabinet a single group or a set of multiple autonomous actors? When a junta is a coordinating body for several parties or groups, is it a single-group or multiple-autonomous-actors decision unit? In other words, what are the rules for deciding which of these two types of decision units has ultimate authority when the entities have characteristics of both kinds? The following discussion builds on our experiences with these questions.

When the occasion for decision is under the purview of a predominant leader who has brought in a set of advisers, we become interested in knowing something about what has happened in the past in such a leader-advisory setting. Does this particular predominant leader tend to use his advisers as consultants to provide him with information and expertise but reserve the right to make the decision himself? Or does this leader see his advisers as forming a team with himself to make decisions on issues like that under consideration? In the first instance, we would have a predominant leader decision unit; in the second instance a single group with a dominant leader. As long as the leader retains the ability to make the choice he or she prefers, the decision unit is a predominant leader. If, however, the leader views the advisers as members of a decision-making team, the decision unit takes on the characteristics of a single group that is interactive and collective in its decision making.

If the decision unit appears to be a coalition cabinet, we consider the following in determining if we have a single-group or multiple-autonomous-actors decision unit. A coalition cabinet is a single group if the dynamics and structures of the parties represented in the coalition do not intrude into the decision-making process. A coalition cabinet is functioning as multiple autonomous actors when the members of the cabinet generally participate in a two-level process in which there is interaction with the leadership of the parties as well as with other members in the cabinet in coping with problems. Four conditions seem relevant to choosing which of these two situations holds. The coalition cabinet is a single group when (1) its members are the heads or leaders (de facto or de jure) of the various parties represented in the cabinet; (2) it has norms through which the parties give cabinet members wide discretion in making decisions; (3) the occasion for decision facing it is time urgent and there is little opportunity to check with the parties; or (4) it generally conducts its deliberations in secret. In each of these instances there is less need, time, and/or reason to check back with the parties. Only those present at the cabinet session participate in the decision-making process. A coalition cabinet is composed of multiple independent actors when the opposite conditions hold; that is, members are not party leaders, norms give members little discretion in making decisions, there is an extended period of time for decision making, or decisions are being made on the record. For such cabinets, the parties become participants in the process as well as their representatives who are members of the cabinets.

Coordinating committees, boards of directors, and interagency groups can pose the same type of dilemma for distinguishing between single-group and coalition decision units. If, for example, a junta is formed from the various types of participants in a coup to coordinate policy for the government, is it a single group or coalition? At issue is the amount of control the coordinating committee can exercise over the multiple actors. If it can make decisions that cannot be readily reversed or modified by outside participants, the coordinators probably form a single-group decision unit. If, however, the outside entities can stipulate what the coordinating committee can and cannot do in response to a given occasion for decision, then we have a coalition decision unit. Some conditions that we have found to coincide with coordinating committees being considered as single groups include the following: (1) the group's existence does not depend on the endorsement of the various autonomous actors outside the group as it is established by law, accepted norms, or long continuous existence; (2) representatives to the coordinating committee cannot be readily replaced by their "home" entity once they are appointed; or (3) representatives to the coordinating committee are all heads of their respective organizations and have considerable substantive discretion in making decisions in the group.

## *Decision Unit Dynamics*

Each kind of authoritative decision unit exists in one of several states that determines the nature of the decision process and the decision calculus for that unit. For each type of decision unit there is a particular "key contingency" that permits us to differentiate configurations leading it to operate in fundamentally different ways. For predominant leader decision units, the individual's sensitivity to information from the political context helps to define how much attention he or she will pay to others' points of view and to situational cues (see M. Hermann, 1984, 1993; Hermann and Kegley, 1995; Kaarbo and Hermann, 1998; Hermann and Preston, 1999). The less sensitive the leader, the more important his or her leadership style and beliefs become in determining what will happen; such leaders are usually more interested in persuading others and in carrying out their own agendas and programs than in seeking advice or listening to others' points of view. These leaders want people around them who will implement their decisions and who will provide them with confirming rather than disconfirming information. Contextual cues are highly relevant to the more sensitive predominant leader who does not take action until the positions of important constituencies and pressures from the domestic and international environments are taken into account. The situation, not the person, becomes the focus of attention.

The operation of a single-group decision unit is, in contrast, conditioned largely by the techniques that are used to manage disagreement and conflict within the group. Since conflict in a problem-solving group can be debilitating, members often devote energy to developing ways of dealing with substantive differences. The literature indicates three different models to describe how single groups cope with conflict: (1) members act to minimize conflict by promoting concurrence ("groupthink"); (2) they acknowledge that disagreement is a fundamental, often unavoidable, part of the decision-making process and seek to resolve the conflict through debate and compromise ("bureaucratic politics"); and (3) even though disagreements are likely, members recognize that such conflict may have no resolution and enact a rule to govern decision making ("winning majority") (see Janis, 1982; C. Hermann, 1993; 't Hart, Stern, and Sundelius, 1997; Stern and Verbeek, 1998; George and George, 1998). Two variables that help to differentiate among these models of group decision making are the extent to which members of the group identify with the group itself or with external organizations and the decision rules used by the group. A focus on building concurrence and denying conflict is more likely when members' loyalties lie within the group. Members are concerned with what is happening in the group itself, in maintaining morale and cohesiveness, and in retaining their position in the group. When members' loyalties lie outside the group, the rules in place to guide decision making help to differentiate if conflict is accepted or resolved. With a unanimity decision rule, members recognize that no solution is possible unless it is acceptable to everyone; with a majority decision rule, members agree to abide by what a certain percentage of the group decides (see Miller, 1989; Levine and Moreland, 1990).

Basic political processes within a coalition of autonomous actors are influenced by the nature of the rules and procedures guiding interaction—the "rules of the game" that shape what is possible (see Hagan, 1993, 1994, 1995; Kaarbo, 1996; Hermann and Hagan, 1998). In particular, we are interested in the degree to which political procedures and norms are well established and "institutionalized." Where rules are essentially absent, we generally find a degree of anarchy and instability in the government with different actors vying for power. When there are established rules, the nature of the rules and theories of coalition formation help us ascertain the decisions that are likely to prevail. If the decision rules permit an authoritative decision when a subset of actors (i.e., a majority) achieves agreement on a course of action, a minimum connected winning coalition is possible. If the decision rules—or political reality—require unanimity among all participants in the coalition, we have a "unit veto" system in which any single actor can block the initiatives of all others.

Knowledge about the nature of these key contingencies provides us with core theoretical insights into the operation of the decision units. Indeed, these key contingencies act as a kind of "theory selector" indicating which models of decision making we need to focus on in understanding the linkage between the decision process and outcome. Table 1 shows how the theory selector works. As the table indicates, each of the basic decision units can be found in three different forms depending on the nature of certain contingencies. The resulting nine types of decision units tap into a wide range of research and theory on how decisions are made by individuals and in groups, organizations, institutions, and political systems. Moreover, the decision units in this elaborated categorization engage in foreign policy decision making with different aims and highlighting different processes. Thus, for example, a coalition with no established rules will probably deadlock since the parties that make up the decision unit are less interested in resolving the substantive problem than in gaining control and power for themselves. A relatively insensitive predominant leader is likely to take strong, forceful actions as he or she moves to put into place an agenda or push for a cause. Members of a single group with an interest in resolving conflict but not wishing to "lose face" with the organizations they represent are likely to want to "paper over differences" or to engage in building a compromise all can accept.

In addition to telling us what theories to apply in understanding a particular decision, the contingencies also provide us with insights concerning whether the decision unit will be "open" or "closed" to the pressures of its environment, both domestic and international. In effect, they tell the researcher when to focus on the decision unit itself in determining the nature of the foreign policy decision and when there is a need to look outside the unit for influences that will shape the decision. Decision units with principled (less contextually sensitive) predominant leaders, single groups with strong internal loyalties, and coalitions with poorly established decision rules have internal dynamics that override external pressures and largely dictate their decision outcomes. Consider as examples Mao's leadership of the cultural revolution and isolation of China, decisions made within the Thatcher cabinet following the Argentinian invasion of the Falkland Islands, and the futile attempts at building a cohesive Iranian foreign policy during the hostage-taking crisis when Khomeini had yet to solidify his control over the government. In each case decisions were driven by *internal* dynamics, be it the personalities of principled leaders, the strong loyalty within a single group, or the severe politics within an unstable coalition. In marked

TABLE 1. Decision Unit Dynamics

| Decision Unit | Key Contingency | Theories Exemplify | Decision Process |
|---|---|---|---|
| Predominant Leader | Sensitivity to Contextual Information: | | |
| | (a) Relatively Insensitive (Goals and Means Well-Defined) | Personality Theory | Principled |
| | (b) Moderately Sensitive (Goals Well-Defined, Means Flexible; Political Timing Important) | Theories Based on the Person/Situation Interaction | Strategic |
| | (c) Highly Sensitive (Goals and Means Flexible) | Theories Focused on the Situation Alone | Pragmatic |
| Single Group | Techniques Used to Manage Conflict in Group: | | |
| | (a) Members Act to Minimize Conflict (Members Loyal to Group) | Group Dynamics ("Groupthink") | Deny Conflict and Seek Concurrence |
| | (b) Members Acknowledge Conflict Is Unavoidable; Group Must Deal with It (Members' Loyalty Outside Group; Unanimity Decision Rule) | Bureaucratic Politics | Resolve Conflict Through Debate and Compromise |
| | (c) Members Recognize Conflict May Have No Resolution (Members' Loyalty Outside Group; Majority Decision Rule) | Minority/Majority Influence and Jury Decision Making | Accept Conflict and Allow for Winning Majority |
| Coalition | Nature of Rules/Norms Guiding Interaction: | | |
| | (a) No Established Rules for Decision Making | Theories of Political Instability | Anarchy |
| | (b) Established Norms Favor Majority Rule | Theories of Coalition Formation | Minimum Connected Winning Coalition |
| | (c) Established Norms Favor Unanimity Rule | Theories Regarding Development of Under- and Over-Sized Coalitions | Unit Veto |

contrast, those decision units characterized by more sensitive predominant leaders, single groups whose members' primary identities are to other entities, and coalitions with well-established rules are penetrable and more susceptible to outside sources of influence; that is, they are "open" and more likely to take into account what is going on in the particular situation both domestically and internationally. As illustrations, note Brezhnev's waffling over the decision to invade Czechoslovakia in 1968 until he sensed there was unanimity among members of the Politburo and their important constituencies in taking such action, the struggles currently going on within the Israeli coalition cabinet as each party tries to gain domestic leverage through slowing down or speeding up the Middle East peace process, and the debate that occurred between President Clinton and the Congress over giving China most favored nation status.

The dynamics that characterize these different types of decision units are described in more detail in the three articles that follow this piece. Each article focuses on one of the basic decision units—predominant leader, single group, and coalition of autonomous actors—and elaborates the ways in which the various contingencies affect what happens in the foreign policymaking process. In addition, we asked foreign policy analysts who had studied particular foreign policy decisions extensively to apply the framework to their cases. These applications of the framework are included in the discussions of each type of decision unit and provide illustrations of how the framework can be used.

## *Outputs of the Decision Units Approach*

There are two types of outputs from the decision units framework. First, there are the outcomes of the decision process itself. What happens when the decision unit configured in a particular way tries to cope with a specific occasion for decision? We call what occurs when the decision unit engages in decision making "process outcomes." In effect, process outcomes denote whose positions have counted in the final decision. Second, there are the actual foreign policy actions that are taken by the government. What is the substantive nature of the decision? In other words, how would we describe what the government, as represented by the particular authoritative decision unit, decided to do in substantive terms in response to an occasion for decision? Thus, one of the outputs records what happened in the decision process; the other indicates the content of the foreign policy decision that resulted from the choice process.

***Process outcomes.*** There appear to be at least six possible outcomes in a decision-making process, that is, six distinctly different things that can happen in the course of a decision unit's deliberations. These six include "one party's position prevails," "concurrence," "mutual compromise/consensus," "lopsided compromise," "deadlock," and "fragmented symbolic action." In each case the outcome of the process indicates the endpoint of the decision in terms of the preferences of those involved. Thus, when one party's position prevails, some of those in the decision process have their preferences accepted as the choice. In concurrence there is a shared sense of direction that either results from the decision process or is evident in the preferences of those involved in the beginning of the process. A mutual compromise/consensus indicates that all parties in the decision unit have yielded some of their position in order, in turn, not to lose out completely in the choice process. A lopsided compromise, in contrast, suggests that one party's preferences have prevailed but they have yielded a little to allow the others in the decision process to save face. With deadlock, those in the decision unit cannot agree and, in effect, at this moment in time "agree to disagree." Fragmented symbolic action is a deadlock in which the disagreement explodes outside the decision unit with each participant in the decision unit trying to take action on their own and/or complaining about the others' behavior. Such activity is often confusing to other actors in the international system who see what appears to be a set of uncoordinated behaviors on the part of the government and wonder who is in charge.

Deadlock, compromise, and concurrence can be arrayed along a dimension that shows how representative the decision is of the range of preferences of those who make up the decision unit. In a deadlock no one's preferences are represented in the decision because the unit is unable to reach a decision. In a compromise everyone gets some of what they want—the partial preferences of everyone are represented in the decision. And in concurrence, the decision represents the shared preferences of everyone.

This dimension has implications for how acceptable the decision will be to participants in the unit and for the extremity of any substantive response that is chosen. There is less closure to the decision process when a deadlock occurs than when the decision unit can reach concurrence. Deadlock leaves the occasion for decision unresolved; no one has won out but nothing has happened to deal with the problem either. Compromise suggests that everything that is possible at the moment with this set of participants has happened but leaves room for revisiting the decision at a later time when the situation or decision unit may have changed. Only in the concurrence situation is there real closure and a shared sense of movement on the problem. The decision represents what all wanted.

As a result, the decision is likely to be more extreme when there is concurrence than when there is compromise or deadlock. In fact, since deadlock usually results in minimal or no action, there is little substance to the foreign policy behavior when the decision process leads to deadlock. Because compromise—particularly a compromise with mutual concessions or one that represents a papering over of differences—indicates a give and take on the part of all participants, we argue that it leads to a more moderated foreign policy behavior.

The commitment of resources, the affective feeling indicated in any action, the instruments used, and the degree of initiative taken are likely to be more middle of the road than extreme with compromise. In effect, participants are adjusting their preferences toward the mean in the process of reaching a compromise. Therefore, decisions arrived at through concurrence are likely to be the most extreme content-wise. Based on a shared sense of what needs to be done in dealing with the problem, participants can take (or not take) initiatives, commit (or refuse to commit) resources, engage in military and economic activities rather than just diplomacy, and be openly cooperative or conflictual. They know that the others are onboard and back what is happening. The process can center on what to do rather than on mediating disagreements about what to do.

In the process outcomes we have just described members of the authoritative decision unit receive symmetrical payoffs. All parties are treated alike. Even in the deadlock where no decision is made, all participants experience the same outcome. Such is not the case, however, for some of the process outcomes. When one party's position prevails, there is a lopsided compromise, or there is a deadlock where the parties do not agree to disagree, the payoffs are asymmetrical—some of the participants benefit in the choice process while others lose out. That is, some members of the decision unit realize their preferences while others do not. Influence on the decision is unequal among the members of the unit.

The symmetrical-asymmetrical quality of process outcomes provides information about the instability of the outcome—the likelihood that the decision unit will want to revisit the decision later on. Some of the outcomes lead to a greater sense of deprivation on the part of those actors who do not "win" or see their preferences realized in the decision. For example, in a deadlock where the participants agree to disagree there is more stability to the outcome than if there is a stalemate among the actors with all leaving the decision unit deliberations still arguing their case and seeking others outside the decision unit to join in the fray. The first we have designated a deadlock; the second we have called fragmented symbolic action because it generally involves all parties taking their own actions with no coordinated governmental response. While the participants in the decision unit in a deadlock have come to closure, when the outcome is fragmented symbolic action the participants are worked up over the process and continue to try to exert power over one another by engaging in activity outside the decision unit.

A similar phenomenon occurs with compromise. The decision unit can arrive at a compromise where there are mutual concessions and all participants believe they have gained something or, at the least, not lost all they could have in the decision process. And the decision process can lead to a lopsided compromise where one party gains more than the other(s). In a lopsided compromise, generally the party that "wins" has offered the other(s) a way to save face. But those who must give more have reason to monitor what happens and urge the decision be reexamined if the negative consequences they expect actually appear to be happening. A comparable case can also be made where one party's position prevails. When one party gets their way, the other actors are deprived of doing what they want to do and, again, have reason to call for a different decision if the desired consequences are not forthcoming. If, however, one party's position prevails because there is concurrence or consensus among the participants or a shared belief in the position selected, everyone's position is represented and the decision can have a certain finality.

Those members of the decision unit whose positions are not represented in the outcome can become agitators for different or further action. They are the part of the decision unit likely to want to keep the issue alive by pushing for reconsideration of the decision, by showing how the previous policy is not achieving what they perceive to be the objective, or by urging that a different or reconstituted decision unit examine the problem. For such members, the occasion for decision remains something of a "cause célèbre," particularly if they were strong advocates of the position that was not chosen. They become part of the shadow of the future and help to determine the implications of that shadow on future decisions. If these members become too vocal, there may be attempts by the rest of the decision unit to reconstitute the unit by isolating them—for example, by bypassing them in future decisions or meeting when they cannot attend. The Iran-contra actions during the Reagan administration illustrate these points. Weinberger and Schulz, respectively U. S. secretaries of defense and state under Reagan, were adamant in their criticisms of trading arms for hostages and urged at several points when key decisions were being made that the National Security Council staff stop their efforts to deal with Iran. As noted in the Tower Commission Report (Tower, Muskie, and Scowcroft, 1987) documenting and evaluating this case, these two policymakers were excluded from the decision process as they became more vocal in their critique of what was going on and had to find other ways of reaching Reagan to argue their case.

Table 2 summarizes our discussion. As is evident in the table, the process outcomes indicate different degrees of ownership of the choice that is made, different ways of monitoring what happens as a consequence of the decision, and different effects on the structure of the decision unit in the future. Indeed, by knowing what the process outcome is, we gain information about what is likely to happen as a result of the decision. A concurrence process outcome is the most likely to move the decision unit to other issues and problems unless the immediate feedback is highly negative. There is a certain finality to the decision until the situation changes or there is some kind of reaction from the international or domestic arena. At the other extreme is a decision process leading to fragmented symbolic action where no one is satisfied and all are working to seek outside support in order to reconstitute the nature of the decision unit

**TABLE 2. Characteristics and Implications of Process Outcomes**

| Range of Preferences Represented in Decision | Distribution of Payoffs | |
| --- | --- | --- |
| | Symmetrical | Asymmetrical |
| One Party's | Concurrence (All own decision; see decision as final; move to other problems) | One Party's Position Prevails (Only one party owns decision; others monitor resulting action; push for reconsideration if feedback negative) |
| Mixed Parties' | Mutual Compromise/ Consensus (Members know got all possible at moment; monitor for change in political context; seek to return to decision if think can change outcome in their favor) | Lopsided Compromise (Some members own position, others do not; latter monitor resulting action and political context, agitating for reconsideration of decision) |
| No Party's | Deadlock (Members know no one did better than others; seek to redefine the problem so solution or trade-offs are feasible) | Fragmented Symbolic Action (No members own decision; seek to change the political context in order to reconstitute decision unit) |

more to their liking and closer to their position. The situation is very fluid with all parties jockeying for position. The rest of the process outcomes fall between these two extremes. When the distribution of payoffs is symmetrical, there is interest in monitoring the political context for change that might favor getting more than the other party(ies) but relations are viewed as okay at the moment. When the distribution is asymmetric, however, at least one party is generally agitated by what happened and ready to see the solution revisited. The reexamination is most pressing where there is fragmented symbolic action. Such pressure is more focused when one party's position prevails or there is a lopsided compromise. With these latter two outcomes, there is a specific target to attack and on which to center opposition.

In Table 1, we observed that the theories of decision making on which the decision units framework is based indicated a particular process outcome was characteristic of a specific type of decision unit. Having now described the set of process outcomes that form the outputs for the decision units approach, it is possible to translate the dominant processes mentioned in the general decision-making literature into one of the six process outcomes that form the outputs of our framework. Consider the following:

Predominant Leader—

- Relatively Insensitive to Political Context—One Party's Position Prevails
- Moderately Sensitive to Political Context—Inaction, Concurrence, or Lopsided Compromise depending on feasibility of preferred option
- Highly Sensitive to Political Context—Mutual Compromise

Single Group—

- Members Act to Minimize Conflict—Concurrence
- Members Acknowledge Conflict Is Unavoidable; Group Must Deal with It—Mutual Compromise or Deadlock
- Members Recognize Conflict May Have No Resolution So Accept Majority Rule—One Party's Position Prevails

Coalition—

- No Established Rules for Decision Making—Fragmented Symbolic Action
- Established Norms Favor Majority Rule—One Party's Position Prevails
- Established Norms Favor Unanimity—Deadlock or Mutual Compromise

The rationale behind these linkages is spelled out in detail in the next three articles in this special issue. Moreover, some techniques that members of these various decision units have used at times to achieve a different process outcome than that posed in the decision-making literature are also elaborated in these three articles.

***Substantive outcomes.*** We are interested in the substantive nature of the decision as well as the process outcome. If we know in general the positions of the members of the decision unit, the process outcomes tell us which of these members' options were taken into consideration in the decision and, thus, provide us with some information about what the decision was. The question, however, becomes, are the process outcomes suggestive of certain types of substantive outcomes?

In another place, the author (Hermann and Hermann, 1989) has examined the relationship between process outcome and the extremity of the foreign policy response. That study built on the observation made earlier that more extreme responses will characterize decisions involving concurrence than where there is deadlock or compromise. In this research the substance of the decisions was defined by the attributes of the foreign policy actions that were taken. Among the attributes included in the study were the degree of commitment of government resources to the activity, the type of statecraft the action required (diplomatic, economic, or military), and the intensity of the feeling, or affect, that accompanied the action chosen. These attributes are discussed in more detail in Callahan, Brady, and Hermann (1982). Examining five thousand decisions for twenty-five countries across a decade, we found support for the hypothesized relationship. Decisions resulting from a concurrence process involved a higher commitment of resources by the government, stronger expressed affect (for both cooperative and conflictual actions), and more of a focus on economic and military instruments of statecraft (on doing something) as opposed to diplomatic (only talking) than occurred when the process outcome was a compromise. Deadlocks led to minimal commitment of resources to the foreign policy activity, rather neutral affect expressed in any discussion of the decision, and a diplomatic response if one was demanded.

This research suggested the linkages between process and substantive outcomes presented in Table 3. We have added a fourth attribute of foreign policy behavior to this table which also seems related to the process outcomes: the degree to which the decision involves an initiative vs. a reaction to something in the international arena. As this table indicates, it is possible to move from the

TABLE 3. Substantive Nature of Decisions Corresponding to Various Process Outcomes

| Process Outcome | Attributes of the Foreign Policy Response | | | |
| --- | --- | --- | --- | --- |
| | Commitment of Resources | Intensity of Affect | Willingness to Take Initiatives | Instruments of Statecraft Used |
| Concurrence | High | Strong | High | Military/Economic |
| One Party's Position Prevails | High/Moderate | Moderate | High/Moderate | Military/Economic |
| Lopsided Compromise | High/Moderate | Moderate/Low | High/Moderate | Diplomatic plus Military/Economic |
| Mutual Compromise | Moderate | Low | Moderate | Diplomatic |
| Fragmented Symbolic Action | Minimal | Strong | Each Actor Takes Initiatives on Own | Diplomatic |
| Deadlock | Minimal | Neutral | React | Diplomatic If Any Response Demanded |

process outcomes to knowing something about the attributes of the resulting foreign policy action. Indeed, the decisions of the units become less extreme as we move down the process outcomes from concurrence to deadlock. The lopsided compromise and fragmented symbolic action result in somewhat more dramatic responses than the casual observer might expect. In fact, the lopsided compromise is more like the decision where one party's position prevails than is the mutual compromise. And with fragmented symbolic action, each member of the decision unit acts as if their particular position had prevailed, taking action on their own even though it does not represent the larger political entity.

The attributes of the foreign policy actions change as we move down Table 3 as well. Commitment of the government's resources is likely to be higher when there is concurrence or one party's preferences are represented in the decision than when the decision represents no one's position or a mixture of the preferences of the parties involved. Similarly, foreign policy initiatives are more likely to be taken when there is concurrence or one party's preferences are represented in the decision, while reactions to stimuli from the international environment are more likely when deadlock is the outcome of the decision process. It is difficult to do more than respond in this latter situation since there is no agreement on what to do or even if any action is necessary. Decisions are less likely to involve the use of a government's higher-priced instruments of statecraft (economic and military) when they involve compromise or deadlock than when the members of the decision unit concur or one party's position prevails. The temporary character of compromises and deadlocks suggests that any decision will be fairly tentative. The feeling tone of the decisions also seems likely to be more intense when one party pushes their position through—particularly if that position is one all agree to—than when there is a compromise or deadlock. The one exception to this norm is where the parties deadlock but cannot even agree to disagree and move to take actions on their own. Then affect becomes more intense in tone. In essence, when the decision represents a compromise or deadlock, foreign policy actions are more constrained than when a single position wins out.

## In Conclusion

It has been the premise of this essay that the decision unit involved in making foreign policy can shape the nature of that policy. Whether the decision unit is a powerful leader, a single group, or a coalition of autonomous actors makes a difference in what governments can do in the international arena. In this article we have presented a framework for understanding how decision units influence the foreign policymaking process that is both derived from and attempts to integrate models of decision making that are extant in the literatures on international relations and foreign policy analysis as well as elsewhere in the social sciences. Although the decision units approach presented here is organized around the notion of three basic decision units, the behavior of the units is contingent on a set of key factors that help to specify the relevance of the immediate international situation and the political context within and outside the government as well as indicate the particular decision process and outcome that are likely to result. In this way, the framework provides the greater theoretical comprehensiveness and flexibility needed to do comparative analyses of foreign policy decision making. Our characterization of the three types of decision units is intended to capture the full array of actors that appear to be involved in foreign policymaking across political systems. In fact, the problem focus and contingency-based logic does not preclude any type of decision unit from occurring in any kind of polity. As a result, the contingencies that are key to determining the nature of the decision unit and process provide us with a mechanism for integrating different, yet complementary, conceptions of who is involved in foreign policymaking and how they are likely to affect the decision that is chosen. Said another way, we have established a blueprint that can be applied to specific cases of foreign policymaking in different countries to see if the proposed accounts of the effects of decision units give an adequate explanation of what really happens.

The rest of the articles in this special issue elaborate the decision units framework that has been described in this piece. As noted earlier, the next three pieces describe the three basic types of decision units, providing more detail on how the key contingencies operate and presenting more specifics about how the units incorporate optimal as well as suboptimal decision processes. Each article includes a set of cases that exemplify how the decision units approach works with that type of decision unit. The authors selected to do the case studies have all done extensive archival research on their cases and, thus, have in-depth knowledge about what did, indeed, happen historically in that particular set of events. After applying the framework, these authors were asked to compare the process outcome they found with what took place in the particular case and to evaluate the framework. We have chosen cases that not only illustrate the three types of decision units but also represent events and countries from around the world. The cases occurred in developing and developed countries as well as in states from various regions of the world.

The last article in this special issue reports a study of the application of the decision units approach to a large bank of cases (some sixty-five) and facilitates a more systematic exploration of the match between the framework and historical reality. In this research, we sought feedback that was intended to help in improving the framework. Thoughtful readers will already want to add features that we have excluded to date from this effort; they should hold onto those considerations until this last article where we discuss ways of amending the framework suggested by the further examination of a large number of problems, occasions for decision, and decisions.

# REFERENCES

ALLISON, GRAHAM T. (1971) *Essence of Decision: Explaining the Cuban Missile Crisis.* Boston: Little, Brown.

ALLISON, GRAHAM T., AND PHILIP ZELIKOW (1999) *Essence of Decision: Explaining the Cuban Missile Crisis*, 2nd ed. New York: Longman.

BARBER, JAMES DAVID (1977) *The Presidential Character.* Englewood Cliffs, NJ: Prentice-Hall.

BENDOR, JONATHAN, AND THOMAS H. HAMMOND (1992) Rethinking Allison's Models. *American Political Science Review* **86**:301–322.

BURKE, JOHN P., AND FRED I. GREENSTEIN (1991) *How Presidents Test Reality: Decisions on Vietnam, 1954 and 1965.* New York: Russell Sage Foundation.

BURNS, JAMES MACGREGOR (1978) *Leadership.* New York: Harper and Row.

CALDWELL, DAN, AND TIMOTHY J. MCKEOWN (1993) *Diplomacy, Force, and Leadership.* Boulder, CO: Westview Press.

CALLAHAN, PATRICK, LINDA P. BRADY, AND MARGARET G. HERMANN (1982) *Describing Foreign Policy Behavior.* Beverly Hills, CA: Sage.

EVANS, PETER, HAROLD JACOBSON, AND ROBERT PUTNAM (1993) *Double-Edged Diplomacy: International Bargaining and Domestic Politics.* Berkeley: University of California Press.

FRANKS, LORD (1983) *Falkland Islands Review.* London: Her Majesty's Stationery Office.

GALVIN, JOHN C. (1994) Breaking Through and Being Heard. *Mershon International Studies Review* **38**:173–174.

GEORGE, ALEXANDER L. (1980) *Presidential Decisionmaking in Foreign Policy: The Effective Use of Information and Advice.* Boulder, CO: Westview Press.

GEORGE, ALEXANDER L. (1993) *Bridging the Gap: Theory and Practice in Foreign Policy.* Washington, DC: United States Institute of Peace.

GEORGE, ALEXANDER L., AND JULIETTE L. GEORGE (1998) *Presidential Personality and Performance.* Boulder, CO: Westview Press.

GREENSTEIN, FRED I. (1987) *Personality and Politics.* Princeton, NJ: Princeton University Press.

GUNTHER, RICHARD (1988) *Politics and Culture in Spain*. Politics and Culture Series no. 5. Ann Arbor: Institute for Social Research, University of Michigan.

HAGAN, JOE D. (1993) *Political Opposition and Foreign Policy in Comparative Perspective*. Boulder, CO: Lynne Rienner.

HAGAN, JOE D. (1994) Domestic Political Systems and War Proneness. *Mershon International Studies Review* **38**:183–207.

HAGAN, JOE D. (1995) "Domestic Political Explanations in the Analysis of Foreign Policy." In *Foreign Policy Analysis*, edited by Laura Neack, Jeanne A. K. Hey, and Patrick J. Haney. Englewood Cliffs, NJ: Prentice-Hall.

HALPERIN, MORTON H. (1974) *Bureaucratic Politics and Foreign Policy*. Washington, DC: Brookings Institution.

HAMPSON, FEN O. (1988) "The Divided Decision Maker: American Domestic Politics and the Cuban Crisis." In *The Domestic Sources of American Foreign Policy*, edited by Charles W. Kegley, Jr., and Eugene R. Wittkopf. New York: St. Martin's Press.

'T HART, PAUL (1990) *Groupthink in Government: A Study of Small Groups and Policy Failure*. Amsterdam: Swets and Zeitlinger.

'T HART, PAUL, ERIC K. STERN, AND BENGT SUNDELIUS (1997) *Beyond Groupthink: Political Group Dynamics and Foreign Policymaking*. Ann Arbor: University of Michigan Press.

HASTINGS, MAX, AND SIMON JENKINS (1983) *The Battle for the Falklands*. New York: Norton.

HERMANN, CHARLES F. (1972) *International Crises: Insights from Behavioral Research*. New York: Free Press.

HERMANN, CHARLES F. (1993) "Avoiding Pathologies in Foreign Policy Decision Groups." In *Diplomacy, Force, and Leadership*, edited by Dan Caldwell and Timothy J. McKeown. Boulder, CO: Westview Press.

HERMANN, CHARLES F., AND MARGARET G. HERMANN (1985) "The Synthetic Role of Decision-Making Models in Theories of Foreign Policy: Bases for a Computer Simulation." In *Simulation and Models in International Relations*, edited by Michael Ward. Boulder, CO: Westview Press.

HERMANN, MARGARET G. (1976) "Circumstances Under Which Leader Personality Will Affect Foreign Policy." In *In Search of Global Patterns*, edited by James N. Rosenau. New York: Free Press.

HERMANN, MARGARET G. (1984) "Personality and Foreign Policy Decision Making: A Study of 53 Heads of Government." In *Foreign Policy Decision Making: Perceptions, Cognition, and Artificial Intelligence*, edited by Donald A. Sylvan and Steve Chan. New York: Praeger.

HERMANN, MARGARET G. (1988) "The Role of Leaders and Leadership in the Making of American Foreign Policy." In *The Domestic Sources of American Foreign Policy*, edited by Charles W. Kegley, Jr., and Eugene R. Wittkopf. New York: St. Martin's Press.

HERMANN, MARGARET G. (1993) "Leaders and Foreign Policy Decision Making." In *Diplomacy, Force, and Leadership*, edited by Dan Caldwell and Timothy J. McKeown. Boulder, CO: Westview Press.

HERMANN, MARGARET G., AND JOE D. HAGAN (1998) International Decision Making: Leadership Matters. *Foreign Policy* **100**(spring):124–137.

HERMANN, MARGARET G., AND CHARLES F. HERMANN (1982) "A Look Inside the Black Box: The Need for a New Area of Inquiry." In *Biopolitics, Political Psychology, and International Politics*, edited by Gerald W. Hopple. New York: St. Martin's Press.

HERMANN, MARGARET G., AND CHARLES F. HERMANN (1989) Who Makes Foreign Policy Decisions and How: An Empirical Inquiry. *International Studies Quarterly* **33**:361–387.

HERMANN, MARGARET G., CHARLES F. HERMANN, AND JOE D. HAGAN (1987) "How Decision Units Shape Foreign Policy Behavior." In *New Directions in the Study of Foreign Policy*, edited by Charles F. Hermann, Charles W. Kegley, Jr., and James N. Rosenau. London: Allen & Unwin.

HERMANN, MARGARET G., AND CHARLES W. KEGLEY, JR. (1995) Rethinking Democracy and International Peace: Perspectives from Political Psychology. *International Studies Quarterly* **39**:511–533.

HERMANN, MARGARET G., AND THOMAS PRESTON (1994) Presidents, Advisers, and Foreign Policy: The Effect of Leadership Style on Executive Arrangements. *Political Psychology* **15**:75–96.

HERMANN, MARGARET G., AND THOMAS PRESTON (1999) "Presidents, Leadership Style, and the Advisory Process." In *Domestic Sources of American Foreign Policy*, edited by James McCormick and Eugene R. Wittkopf. Lanham, MD: Rowman and Littlefield.

HOLSTI, OLE R. (1976) "Foreign Policy Formation Viewed Cognitively." In *The Structure of Decision: The Cognitive Maps of Political Elites*, edited by Robert Axelrod. Princeton, NJ: Princeton University Press.

HUDSON, VALERIE M. (1995) Foreign Policy Analysis Yesterday, Today, and Tomorrow. *Mershon International Studies Review* **39**:209–238.

JANIS, IRVING L. (1982) *Groupthink: Psychological Studies of Policy Decisions and Fiascoes*, 2nd ed. Boston: Houghton Mifflin.

KAARBO, JULIET (1996) Power and Influence in Foreign Policy Decision Making: The Role of Junior Coalition Partners in German and Israeli Foreign Policy. *International Studies Quarterly* **40**:501–530.

KAARBO, JULIET, AND MARGARET G. HERMANN (1998) Leadership Styles of Prime Ministers: How Individual Differences Affect the Foreign Policymaking Process. *Leadership Quarterly* **9**:243–263.

KHONG, YUEN FOONG (1992) *Analogies at War*. Princeton, NJ: Princeton University Press.

KLEIBOER, MARIEKE A. (1998) *International Mediation: The Multiple Realities of Third-Party Intervention*. Boulder, CO: Lynne Rienner.

KORANY, BAGHAT, AND ALI E. HILLAL DESSOUKI (1991) *The Foreign Policies of Arab States: The Challenge of Change*, 2nd ed. Boulder, CO: Westview Press.

KRUZEL, JOSEPH (1994) More a Chasm Than a Gap, But Do Scholars Want to Bridge It? *Mershon International Studies Review* **38**:179–181.

KUPCHAN, CHARLES (1994) *The Vulnerability of Empire*. Ithaca, NY: Cornell University Press.

LEBOW, RICHARD NED (1981) *Between Peace and War: The Nature of International Crisis*. Baltimore, MD: Johns Hopkins University Press.

LEBOW, RICHARD NED (1985) "Miscalculation in the South Atlantic: The Origins of the Falklands War." In *Psychology and Deterrence*, edited by Robert Jervis, Richard Ned Lebow, and Janice Gross Stein. Baltimore, MD: Johns Hopkins University Press.

LEVINE, JOHN M., AND RICHARD L. MORELAND (1990) Progress in Small Group Research. *Annual Review of Psychology* **41**:585–634.

LINCOLN, JENNIE K., AND ELIZABETH G. FERRIS (1984) *The Dynamics of Latin American Foreign Policies: Challenges for the 1980s*. Boulder, CO: Westview Press.

MAOZ, ZEEV (1990) Framing the National Interest: The Manipulation of Foreign Policy Decisions in Group Settings. *World Politics* **43**:77–111.

MILLER, C. E. (1989) "The Social Psychological Effects of Group Decision Rules." In *Psychology of Group Influence*, 2nd ed., edited by P. B. Paulus. Hillsdale, NJ: Erlbaum.

PRESTON, THOMAS (2001) *The President and His Inner Circle: Leadership Style and the Advisory Process in Foreign Policy Making*. New York: Columbia University Press.

ROSATI, JEREL A. (2000) The Power of Human Cognition in the Study of World Politics. *International Studies Review* **2**:45–75.

SNYDER, JACK (1991) *Myths of Empire: Domestic Politics and International Ambition*. Ithaca, NY: Cornell University Press.

STERN, ERIC K. (1999) *Crisis Decisionmaking: A Cognitive-Institutional Approach*. Stockholm: Department of Political Science, Stockholm University.

STERN, ERIC K., AND BERTJAN VERBEEK (1998) Whither the Study of Governmental Politics in Foreign Policymaking? A Symposium. *Mershon International Studies Review* **42**:205–255.

STEWART, PHILIP D., MARGARET G. HERMANN, AND CHARLES F. HERMANN (1989) Modeling the 1973 Soviet Decision to Support Egypt. *American Political Science Review* **83**:35–59.

SYLVAN, DONALD A., AND JAMES F. VOSS (1998) *Problem Representation in Foreign Policy Decision Making*. Cambridge: Cambridge University Press.

TOWER, JOHN, EDMUND MUSKIE, AND BRENT SCOWCROFT (1987) *The Tower Commission Report*. New York: Bantam Books.

VERTZBERGER, YAACOV Y. (1984) Bureaucratic-Organizational Politics and Information Processing in a Developing State. *International Studies Quarterly* **28**:69–95.

VERTZBERGER, YAACOV Y. (1990) *The World in Their Minds: Information Processing, Cognition, and Perception in Foreign Policy Decisionmaking*. Stanford, CA: Stanford University Press.

WEINSTEIN, FRANKLIN B. (1972) The Uses of Foreign Policy in Indonesia: An Approach to the Analysis of Foreign Policy in the Less Developed Countries. *World Politics* **24**:356–381.

WELCH, DAVID A. (1992) The Organizational Process and Bureaucratic Politics Paradigms: Retrospect and Prospect. *International Security* **17**:112–146.

82

# Who Leads Matters:

## The Effects of Powerful Individuals

*Margaret G. Hermann*
Maxwell School, Syracuse University

*Thomas Preston*
Washington State University

*Baghat Korany*
American University, Cairo

*Timothy M. Shaw*
Dalhousie University

An examination of how governments and ruling parties make foreign policy decisions suggests that authority is exercised by three types of decision units: leaders, groups, and coalitions. Moreover, the literature indicates that within any one government the pertinent decision unit often changes with time and issue. In this article we are interested in exploring what happens

---

NOTE: Hermann and Preston are responsible for the theoretical discussion presented here; all four authors applied the theoretical framework to particular leaders and situations which they had studied extensively in other research to explore its applicability. The authors would like to express their appreciation to Michael Young for developing a software program that facilitates assessing leadership style from interviews and speeches with leaders (see www.socialscienceautomation.com) and to Eric Stern and Deborah Wituski for their constructive comments on an earlier draft.

when the decision unit is a single, powerful individual. When such an individual takes responsibility for making the choice regarding how to deal with a foreign policy problem, what effect can he or she have on the substance of the action selected? This type of decision unit is considered to involve a "predominant leader" because one person has the ability to commit the resources of the society and, with respect to the specific problem being confronted, the power to make a decision that cannot be readily reversed.

The focus of attention here is on the importance of leadership style in understanding what predominant leaders will do in formulating foreign policy—on how different ways of dealing with political constraints, processing information, and assuming authority can promote different reactions to what is essentially the same decision-making environment. In what follows we explore the conditions under which the authoritative decision unit is likely to be a predominant leader, the characteristics of such leaders that can shape what they urge on their governments, and the nature of the impact on policy. Throughout we will provide examples of situations when predominant leaders with various leadership styles have acted as decision units and indicate what happened as a means of illustrating the application of the framework we are advancing in these pages.

## THE LEADER AS AUTHORITATIVE DECISION UNIT

When a single individual has the power to make the choice concerning how a state is going to respond to a foreign policy problem, he or she becomes the decision unit and acts as a predominant leader. Under such conditions, once the leader's position is known, those with different points of view generally stop public expression of their own alternative positions out of respect for the leader or fear of reprisals. If these others are allowed to continue discussing additional options, their opinions are no longer relevant to the political outcome of the moment. As Abraham Lincoln is reported to have said to his cabinet: "Gentlemen, the vote is 11 to 1 and the 1 has it." Only Lincoln's vote mattered in this instance; he was acting as a predominant leader and making the authoritative decision.

The decision unit dealing with a particular foreign policy problem is likely to be a predominant leader if the regime has one individual in its leadership who is vested with the authority—by a constitution, law, or general practice—to commit or withhold the resources of the government with regard to the making of foreign policy. A monarch is an illustration of this kind of predominant leader—for example, King Fahd of Saudi Arabia and King Abdullah of Jordan. The decision unit can also be a predominant leader if the foreign policy machinery of the government is organized hierarchically and one person is located at the top of the hierarchy who is accountable for any decisions that are made. As

Harry Truman said about the American presidency: "The buck stops here." Moreover, if a single individual has control over the various forms of coercion available in the society and, as a result, wields power over others, the decision unit can be a predominant leader. Dictatorships and authoritarian regimes often fall into this category and have predominant leaders dealing with foreign policy matters—for instance, Fidel Castro in Cuba, Saddam Hussein in Iraq, and Kim Il Sung in North Korea (see Korany, 1986b).

Even though a political regime has a single, powerful individual who would qualify under the above definition as a predominant leader, that person *must exercise* authority in dealing with the problem under consideration to become the authoritative decision unit. Otherwise, another type of decision unit assumes responsibility for making the decision. Single, powerful leaders have been found to act as predominant leaders under the following conditions (see, e.g., Hermann, 1984, 1988a, 1995; Greenstein, 1987; Hermann and Hermann, 1989; Preston, 2001): (1) they have a general, active interest in, as well as involvement with, foreign and defense issues; (2) the immediate foreign policy problem is perceived by the regime leadership to be critical to the well-being of the regime—it is perceived to be a crisis; (3) the current situation involves high-level diplomacy or protocol (a state visit, a summit meeting, international negotiations); or (4) the issue under consideration is of special interest or concern to the leader. When a single, powerful leader is interested in foreign policy, he or she generally seeks to control the foreign policy agenda and shape what happens. Whether or not they are interested, however, such leaders can be drawn into the formation of foreign policy when their governments are faced with a crisis or they are involved in a summit meeting.

Franco of Spain is an example. Although qualified as a predominant leader given the structure of power in Spain during his tenure, Franco is reported to have had little interest in foreign affairs and to have left much of the foreign policymaking to his foreign and economics ministers. Only when an issue became critical to his regime, such as renegotiation of agreements concerning the American bases in Spain, did he assume the role of predominant leader in the foreign policymaking process (see Gunther, 1988).

The opposite case can hold as well. Even leaders who generally do not have the authority to commit the resources of their governments without consulting with others can act like predominant leaders under certain conditions. When such leaders have an intense interest in foreign affairs or a particular substantive foreign policy issue or find themselves in the midst of an international crisis, they can assume more authority than is ascribed to their positions. Indeed, a number of scholars (e.g., Hermann, 1972; George, 1980; Lebow, 1981; Hampson, 1988; 't Hart, 1990) have observed that in international crises there is a strong tendency for a contraction of authority to the highest levels of government which, even in democracies, decreases usual institutional and normative restraints and increases leaders' decision latitude while at the same time encouraging them to act on their perceptions of the national interest and their images of the public's preferences. This phenomenon has led some to "question the extent to which the foreign policy process of democracies differs from that of autocracies" under crisis conditions (Merritt and Zinnes, 1991:227). Consider Margaret Thatcher's consolidation of her authority when faced with the invasion of the Falkland Islands (Freedman, 1997) and George H. W. Bush's "personal involvement in the diplomacy required to maintain the coalition" during the Gulf War (Crabb and Mulcahy, 1995:254; see also Woodward, 1991; Preston, 2001). In these instances, though not constitutionally or legally designated as predominant leaders, Thatcher and Bush assumed such a role.

## LEADERSHIP STYLE AND FOREIGN POLICYMAKING PRACTICES

In reaction to the historical debate about whether leaders are born with certain leadership propensities or rise in response to the challenges of their times, researchers have uncovered ample instances of individuals who fall into both categories. This result permits meaningful typification and has implications for the foreign policymaking process when the decision unit is a predominant leader. In the study of political leadership, the more familiar categorizations based on this distinction are crusader vs. pragmatist (see, e.g., Stoessinger, 1979; Nixon, 1982), ideologue vs. opportunist (see, e.g., Ziller, 1973; Ziller et al., 1977; Suedfeld, 1992), directive vs. consultative (see, e.g., Lewin and Lippit, 1938; Bass and Valenzi, 1974; Bennis and Nanus, 1985), task-oriented vs. relations-oriented (see, e.g., Byars, 1973; Fiedler and Chemers, 1984; Chemers, 1997), and transformational vs. transactional (see, e.g., Burns, 1978; Bass, 1985, 1997; Hargrove, 1989; Glad, 1996). Regardless of theoretical purpose, these typologies rest on the assumption that the leadership style of one type of leader is guided by a set of ideas, a cause, a problem to be solved, or an ideology, while the leadership style of the other type arises out of the nature of the leadership context or setting in which the leader finds him or herself. As Snyder (1987:202) has observed, one type is more goal driven; the other is more situationally responsive. The differences between these two leadership styles appear to result from the leaders' images of themselves and their perceptions of where their behavior is validated and are suggestive of how sensitive the leaders are likely to be to the current political context (see Hermann and Hermann, 1989; Hermann, 1993).

The more goal-driven leaders—the crusaders, the ideologues, those who are directive, task-oriented, or transformational in focus—interpret the environment through a lens that is structured by their beliefs, attitudes, motives, and passions. They live by the maxim "unto thine own self by true," their sense of self being determined by the congruence between who they are and what they do. As Gardner (1983) has noted, these leaders have an "inside looking outward" perspective on life. They act on the basis of a set of personal standards and seek out leadership positions where their standards generally are reinforced (Browning and Jacob, 1971; DiRenzo, 1977; Hall and Van Houweling, 1995). Because they tend to selectively perceive information from their environment, such leaders have difficulty changing their attitudes and beliefs (Kotter and Lawrence, 1974; Hermann, 1984; Freidman, 1994). Changes that do occur are usually at the margins. Moreover, they choose associates who define issues as they do and who generally share their goals. These leaders value loyalty and often move to shape norms and institutions to facilitate their personal goals (Hermann and Preston, 1994; Preston, 2001).

Leaders who are more responsive to the current situation—the pragmatists, the opportunists, and those who are consultative, relations-oriented, or transactional—tend, to paraphrase Shakespeare, to see life as a theater where there are many roles to be played. Indeed, people are essentially performers whose main function is choosing the "correct" identity for the situation at hand (Goffman, 1959). Such leaders perceive themselves to be flexible and open-minded. They seek to tailor their behavior to fit the demands of the situation in which they find themselves, and, before making a decision, ascertain where others stand with regard to an issue and estimate how various groups and institutions are likely to act (Driver, 1977; Stewart, Hermann, and Hermann, 1989). In essence, the self-image of these leaders is defined by the expectations and interests of others. To become acceptable, ideas, attitudes, beliefs, motives, and passions must receive external validation from relevant others. Contextually responsive leaders seek to maintain extensive information-gathering networks to alert them to changes in the interests and views of important constituencies (Manley, 1969; Hermann, 1988b; Suedfeld and Wallace, 1995). Moreover, they recruit associates who have access to those constituencies on whom their political support depends (Kotter and Lawrence, 1974; Stein, 1994; Preston, 2001).

Research on the foreign policy behavior of governments led by predominant leaders with these two styles (see, e.g., Driver, 1977; Hermann, 1980, 1984; Hermann and Hermann, 1989; Stewart, Hermann, and Hermann, 1989; Snare, 1992) indicates that there are differences in the kinds of actions that each type advocates. The more contextually responsive predominant leaders appear more constrained by the specific domestic settings in which they find themselves than do their more goal-driven counterparts, and, accordingly, are relatively incremental in the activities they urge on their governments. They are less likely to engage in conflict than the predominant leaders who are more goal-driven, and are averse to committing their country's resources to bellicose actions unless the choice enjoys the support of important constituencies. The contextually responsive leaders are predisposed to seek support for their international decisions. Interested in consensus-building and multilateral approaches to foreign policy, they are most comfortable working within the range of permissible choices that their constituents authorize. They are not high risk takers—only if they can mobilize the constituents they perceive are needed to support a particular activity are they likely to move forward. Indeed, such leaders are less likely to pursue extreme policies of any kind (neither confrontation and war nor peace initiatives and international agreements) unless pushed to do so. Contextually responsive predominant leaders are more likely to be led into conflict or cooperative actions than to lobby for their initiation.

Unlike their contextually responsive counterparts, goal-driven predominant leaders come to foreign policy problems with a particular perspective or set of policy priorities. Such leaders reinterpret and redesign situations, their goals and principles defining what is important in foreign policy. Specific issues—economic decline, military security, internal famine, Arab nationalism, illegal immigration—shape these leaders' views concerning their external priorities and their postures toward other actors. Constraints are things to be overcome or dealt with, not accepted; they are obstacles in the way but are viewed as not insurmountable. Indeed, such leaders are not averse to using diversionary tactics (scapegoating, "bashing" the enemy) to "rally constituencies around the flag" thus reducing the effectiveness of domestic opposition that may disagree with a particular action or activity. Goal-driven predominant leaders energetically try various maneuvers to pull policymaking totally under their direction. As a result, they believe they know more about what is happening in foreign policy in their government and can exercise greater control over it (see Hermann and Preston, 1994; Kissinger, 1994; Kaarbo and Hermann, 1998).

Several examples may help to illustrate the relevance of this difference in degree of sensitivity to the political context to understanding the foreign policy decision making of predominant leaders. Consider two leaders who were forced by circumstance to become predominant: Romulo Betancourt of Venezuela and Eisaku Sato of Japan. Betancourt is an example of a goal-driven leader. He was president of Venezuela during the early 1960s and spent much of his tenure in office trying to maintain Venezuela in the democratic community after a decade of military dictatorship. He believed strongly in the importance of having democratic regimes in Latin America and based his foreign policy on his perceptions of the political systems of the countries he viewed as the sources of his problems. If he perceived a state to be a democracy, it was a friend; if a dictatorship, it was a foe. He enunciated this formula into the Betancourt Doctrine and reacted to all governments based on this kind of analysis (Alexander, 1982). Betancourt's beliefs structured the foreign policy behavior he urged on his state.

Sato, prime minister of Japan in the late 1960s and early 1970s, is an example of a contextually responsive leader. Destler and his colleagues (1979:40), in their analysis of the extended textile wrangle that took place between the United States and Japan during this time period, observed that Sato "was an extraordinarily cautious and discreet man. He would 'tap his way across a stone bridge to be sure it was safe'; he avoided making commitments to one position or another, particularly on controversial policy issues, until a general consensus emerged among the influential groups concerned." Sato wanted to know where others stood and what position would garner the most support before he acted. Cues from the situation were important in structuring the decisions he would urge on his government.

As this discussion suggests, by ascertaining how sensitive a predominant leader is to contextual information, we learn where to look for an explanation of what actions and policies such leaders are likely to encourage their governments to take. If the leaders are contextually responsive (i.e., sensitive to contextual information), their behavior will be more pragmatic and situationally driven; we need to examine the particular political problem and setting closely to determine what is likely to happen. Public opinion, the media, the considerations of powerful legislators, potent interest groups, and advisers may all play some role in shaping foreign policy. If the leaders are goal-driven (i.e., less sensitive to contextual information), their behavior will be more focused around their own beliefs, attitudes, passions, and principles. By learning what motivates these leaders we can understand what the governments' foreign policy actions will probably entail. Knowledge about the political setting becomes less important while information about the leader's policy priorities becomes critical. Thus, degree of sensitivity to the political context is a key variable in determining how a predominant leader is going to respond when he or she becomes the authoritative decision unit in the formation of foreign policy.[1] It is the starting point for differentiating among predominant leaders and the processes they will probably use in dealing with foreign policy problems.

## ASSESSING SENSITIVITY TO THE POLITICAL CONTEXT

How do we decide if a predominant leader is more goal-driven or contextually responsive, or, in other words, how sensitive he or she is likely to be to the political setting in working on foreign policy problems? To assess a leader's sensitivity to contextual information, we seek the answers to three questions: (1) How do leaders react to political constraints in their environment—do they

---

[1] This variable is consistent with several others in the political and social psychology literatures, namely, self-monitoring (Snyder, 1987), need for cognition (Petty and Cacioppo, 1986), and integrative complexity (Suedfeld, Tetlock, and Streufert, 1992).

respect or challenge such constraints? (2) How open are leaders to incoming information—do they selectively use information or are they open to information directing their response? (3) What motivates leaders to take action—are they driven by an internal focus of attention or by responses from salient constituents? These questions represent different ways of being sensitive to the political context and are featured prominently in research on how leaders make decisions. The answers to these queries suggest the strategies and leadership styles predominant leaders are likely to use in addressing a foreign policy problem. Let us explore these questions in more detail.

## *Reaction to Political Constraints*

There is much discussion in research on foreign policymaking about the constraints under which leaders must operate in the decision-making process. Indeed, some argue that domestic and international constraints are such that it is more parsimonious to leave leaders out of the explanatory equation altogether (see Hermann and Hermann, 1982; Greenstein, 1987; Hermann and Hagan, 1998; Young and Schafer, 1998; and Byman and Pollack, 2001 for reviews of this debate). But as scholars have moved to consider how domestic and international constraints can interact in shaping foreign policy in the so-called two-level game, they have reinserted the leader as the negotiator who maneuvers the government and state toward some foreign policy action (see, e.g., Putnam, 1988; Evans, Jacobson, and Putnam, 1993). Leaders are viewed as playing a pivotal role in the bargaining that is required to build a consensus with both their domestic constituents and their international counterparts around a particular option. Moreover, as we observed earlier, those interested in organizational and bureaucratic politics have discovered that in situations of high salience to a government there is a contraction of authority to those individuals with ultimate responsibility for the decision. At such times, leaders are generally freed from the usual constraints on their choices. And others have argued that domestic forces have an impact on foreign policy through leaders' strategies for dealing with opposition (see, e.g., Levy, 1989; Snyder, 1991; Hagan, 1994, 1995; Kupchan, 1994). Leaders can use foreign policy to divert attention away from an opposition, to accommodate to the opposition, or to co-opt the opposition's position; each strategy influences the character of the decision. Thus, there are a number of ways in which leaders can become active in dealing with the political constraints in their environments that, in turn, can shape what happens in foreign policy.

We are interested here in how important it is to a particular predominant leader to exert control and influence over the political environment and the constraints that environment poses as opposed to being adaptable to each specific situation and remaining open to responding to the current demands of domestic and international constituencies and circumstances. In other words, is a predominant leader predisposed to be a constraint challenger or constraint respecter? Our previous description of the variable, sensitivity to the political context, suggests that predominant leaders whose leadership style makes them responsive to contextual information are likely to both pay attention to political constraints and work within such parameters. Predominant leaders who are more goal-driven are less likely to perceive political constraints, but if they do they will view them as something to be tested and overcome, not acceded to. Consider the following.

Research has shown that leaders who are relatively insensitive to the context are more intent on meeting a situation head on, achieving quick resolution to problems they are facing, being decisive, and dealing with what is perceived as the problem of the moment (see, e.g., Driver, 1977; Hermann, 1984; Tetlock, 1991; Suedfeld, 1992; Kowert and Hermann, 1997). Their beliefs, attitudes, and passions are highly predictive of their responses to events. Constraints are obstacles but not insurmountable. To facilitate maintaining influence over events, such leaders work to bring policymaking under their control (see, e.g., Hermann and Preston, 1994, 1999; Preston, 2001).

Leaders who are more sensitive to the context have been found to be more (1) empathetic to their surroundings, (2) interested in how relevant constituencies are viewing events and in seeking their support, (3) open to bargaining, trade-offs, and compromise, and (4) focused on events on a case-by-case basis (see, e.g., Driver, 1977; Ziller et al., 1977; Hermann, 1984; Snyder, 1987; Hermann and Hermann, 1989; Tetlock, 1991; Kaarbo and Hermann, 1998). They need support from their political environment before making a decision. Constraints set the parameters for action. Flexibility, political timing, and consensus-building are viewed as important leadership tools.

Although several studies have found that leaders who are less sensitive to the political context tend to come to office in autocratic political systems and those who are more sensitive in democratic political systems (see Hermann, 1984; Hermann and Hermann, 1989), the relationship is not monotonic. Indeed, the correlation between regime type and leaders' scores on a measure of sensitivity to the political context for 110 heads of state in office 1959–1987 was .56 (gamma). But the data are suggestive of how these two types of leaders are likely to deal with constraints. The goal-driven (less sensitive) leaders are going to be more comfortable in a setting where they are in control and able to set the criteria for action, while the contextually responsive (more sensitive) leaders will have increased tolerance for the sharing of power and the time involved in gaining the consent of the governed.

## Openness to Information

In examining the foreign policymaking of American presidents, George (1980; see also Johnson, 1974; Campbell, 1986; Crabb and Mulcahy, 1986; Burke and

Greenstein, 1991; George and Stern, 1998; Preston, 2001) has observed that the kinds of information they wanted in making a decision were shaped by whether they came with a well-formulated vision or agenda framing how data were perceived and interpreted or were interested in studying the situation before choosing a response. These two approaches to information processing not only affected the kinds of data presidents sought but also the types of advisers they wanted around them. A president with an agenda seeks information that reinforces a particular point of view and advisers who are supportive of these predispositions. A president focused on what is happening politically in the current situation is interested in what is "doable" and feasible at this particular point in time and which advisers are experts or highly attuned to important constituencies and can provide insights into the political context and problem of the moment.

Leaders who are less sensitive to the political context act more like advocates. They are intent on finding that information in the environment that supports their definition of the situation or position and overlooking any evidence that may be disconfirmatory. Their attention is focused on persuading others of their positions. As Stewart, Hermann, and Hermann (1989) found in studying decisions of the Soviet Politburo, those leaders who were advocates for a position used their time to build a case and lobby others to their side; they spent little time assessing the nature of the terrain and others' positions.

Leaders who are more sensitive to the political context are, in fact, cuetakers. They both define the problem and identify a position by checking what important others are advocating and doing. Such leaders are interested in information that is both discrepant and supportive of the options on the table at the moment. In the Politburo study referred to above, the leaders who were more sensitive spent time gathering information and talking with salient persons, seeking political insights into who was supporting which option and with what degree of intensity. Such information was important to them in formulating a representation of the problem and selecting a position.

Less sensitive leaders act like the classic cognitive misers from the information processing literature and the more sensitive leaders like the naive scientists or hypothesis testers that are also described in this research (see, e.g., Axelrod, 1976; Ajzen and Fishbein, 1980; Jonsson, 1982; Fiske and Taylor, 1984; Fazio, 1986; Lau and Sears, 1986; Suedfeld, 1992). The cognitive miser's attention "to various aspects of their environment is narrowly focused" and is guided by schemas or images that define the nature of reality (Lau and Sears, 1986:149). They rely on simple rules or heuristics in making a choice, engaging in top-down information processing in which information is sought to maintain or strengthen the original schema. These leaders start "with the conceptualization of what might be present and then look for confirming evidence biasing the processing mechanism to give the expected result" (Lindsay and Norman, 1977:13).

Hypothesis testers, like the more sensitive leaders, engage in bottom-up information processing; rather than imposing structure on the data, they are guided by the evidence they are receiving from the environment. They are likened to naive scientists who seek to learn if their initial reactions to a problem are supported by the facts, or to use information from the environment to develop a position. In other words, such leaders consider what among a range of alternative scenarios is possible in the current context. As pundits say, the leader "runs an idea up the flagpole to see who salutes it." Feedback becomes critical in helping such leaders modify their behavior to fit the situation (see Steinbruner, 1974).

Contextually responsive predominant leaders are likely to be hypothesis testers or cue-takers in response to foreign policy problems, seeking out information from the political setting before urging an action; they will be relatively open to incoming information. Goal-driven predominant leaders are advocates and cognitive misers pushing their agendas and using their vision of the way things should be to tailor information; they will see what they want to see and, thus, will be relatively closed to the range of information that is available.

## *Motivation for Action*

As Barber (1977:8) has argued, leaders' motivation defines the way they "orient [themselves] toward life—not for the moment, but enduringly." It shapes their character, what is important in their lives, and drives them to action. A survey of the literature exploring motivation in political leaders suggests that a variety of needs and incentives push persons into assuming leadership positions in politics (see, e.g., Barber, 1965; Woshinsky, 1973; McClelland, 1975; Walker, 1983; Payne et al., 1984; Snare, 1992; Winter, 1992, 1995). Examination of the resulting list, however, indicates that political leaders are motivated, in general, either by an internal focus—a particular problem or cause, an ideology, a specific set of interests—or by the desire for a certain kind of feedback from those in their environment—seeking acceptance, approval, power, support, status, or acclaim. In one case, what motivates them is internal; they are pushed to act by ideas and images they believe and advocate. In the other instance, they seek a certain kind of relationship with important others and are pulled by forces outside themselves to action. Those leaders focused on problems and causes are less sensitive to the political context; they know what needs doing and do it. The leaders interested in building relationships are more sensitive to the political context because it is only through interaction with others that they can be satisfied and fulfilled.

This difference in motivation is reflective of the two functions leaders have been found to perform in groups, organizations, and institutions: assuring institutional survival (group maintenance, relationship-building) and policy achievement ("getting things done," task performance) (see, e.g., Fiedler, 1967, 1993; Vroom and Yetton, 1973; Bass, 1981; McGrath, 1984; Campbell, 1986; Hargrove, 1989). Choosing one or the other of these two foci of attention produces a particular style of leadership. A focus on building relationships emphasizes interest in the development of consensus, networks, collegial interactions, and the empowerment of others along with heightened attention to interpersonal and social skills as well as attention to image maintenance. A focus on accomplishing something attaches importance to leaders' problem-solving and management skills and interest in agenda-setting, advocacy, and implementation. For those motivated by relationships with others, persuasion and marketing are central to achieving what they want, whereas for those for whom dealing with a cause or solving a problem is highly salient, mobilization and effectiveness feature prominently in movement toward their goals. Again, one type of motivation necessitates more sensitivity to the political context than the other. Building relationships is only possible if there is some sensitivity to what is going on with important others; it is easier for the leader to push to accomplish things without taking much contextual information into account.

## How Leaders Can Matter

Answers to the above questions about how sensitive a leader is likely to be to the political context combine to provide the analyst with information about that individual's leadership style and some clues about the kind of foreign policy behavior he or she will urge on the government when in the role of a predominant leader.[2] Knowledge about how leaders react to constraints, process information, and are motivated to deal with their political environments indicates that there are a wider array of leadership styles than the two that dominate the leadership literature. Table 1 displays the eight leadership styles that result when these three factors are interrelated.

---

[2] One of the authors (Hermann, 1999a) has developed a way of assessing the answers to the three questions posed here from leaders' responses to questions in press conferences and interviews focusing on seven traits that previous research has shown are linked to particular leadership styles. The manual describing this technique is available on the web at www.socialscienceautomation.com along with several examples of applications of the assessment-at-a-distance method to current leaders. Data on 150 heads of state and national leaders from around the world are now available using this technique. A software program, Profiler+, has been developed to do the assessment automatically from machine-readable text.

TABLE 1. **Leadership Style as a Function of Responsiveness to Constraints, Openness to Information, and Motivation**

| Responsiveness to Constraints | Openness to Information | Motivation | |
|---|---|---|---|
| | | Problem Focus | Relationship Focus |
| Challenges Constraints (Becomes a crusader) | Closed to Information | *Expansionistic* (Focus is on expanding one's power and influence) | *Evangelistic* (Focus is on persuading others to accept one's message and join one's cause) |
| Challenges Constraints (Is generally strategic) | Open to Information | *Incremental* (Focus is on maintaining one's maneuverability and flexibility while avoiding the obstacles that continually try to limit both) | *Charismatic* (Focus is on achieving one's agenda by engaging others in the process and persuading them to act) |
| Respects Constraints (Inclined toward pragmatism) | Closed to Information | *Directive* (Focus is on personally guiding policy along paths consistent with one's own views while still working within the norms and rules of one's current position) | *Consultative* (Focus is on monitoring that important others will support, or not actively oppose, what one wants to do in a particular situation) |
| Respects Constraints (Is usually opportunistic) | Open to Information | *Reactive* (Focus is on assessing what is possible in the current situation given the nature of the problem and considering what important constituencies will allow) | *Accommodative* (Focus is on reconciling differences and building consensus, empowering others and sharing accountability in the process) |

## *Crusaders, Strategists, Pragmatists, and Opportunists*

Leaders' methods of dealing with political constraints and information interact to form four ways often used in the media and leading policy journals to describe politicians' leadership styles. They are engaging in a crusade, being strategic, acting pragmatically, or being opportunistic. The leaders are responding differently to their political environments and are being differentially sensitive to the political context. Consider the following.

Those leaders who challenge constraints and are relatively closed to information from the environment (cognitive misers) are the least sensitive to the political contexts in which they find themselves. They are, indeed, usually crusading for or advocating a position and being proactive. If the political context facilitates what such leaders want to do, they can be effective in mobilizing others to action. But "crusaders" do not wait to take action until the time is right. They are like a dog with a bone—they will find a way! By being convinced that available information supports their position, they can often create a very persuasive rationale for what they are doing that gives their actions credibility and legitimacy. Thus, in the decision-making process, such leaders' positions are likely to prevail as they take charge and work to control what happens. Fidel Castro is an example of a crusader. He has spent much of his political career engaged in trying to export the socialist revolution in Latin America and Africa; he has challenged constraints, interpreted events according to his design, and pursued his position religiously from sending guerrilla troops and revolutionaries to providing medical and technical aid to particular grassroots politicians and movements (see Geyer, 1988).

The opposite of the crusaders are the opportunists—those who respect constraints and are open to information from the political setting (hypothesis testers/ naive scientists). For them, knowledge about the political context is crucial; they are the most sensitive to contextual information. Such leaders are expedient, defining the problem and taking a position based on what important others seem to be pushing. Bargaining lies at the heart of the political game; unless some kind of consensus can be built, inaction is preferable to an action that has the potential of losing support and building opposition. Politics is the art of the possible in the current setting and time. A leader on the contemporary political scene with this leadership style is Zoran Djindjic, president of Serbia (Hermann, 1999d). In working across the past decade to remove Slobodan Milosevic from power, Djindjic has acted as a broker and intermediary convening various political groups in an effort to find one that could achieve the goal; he has been willing to move as slowly or quickly as the situation permits. Much like a chameleon, Djindjic has adjusted his behavior to match the situation.

Those leaders who exhibit signs of being both low and high in their sensitivity to the political context—those who challenge constraints but are open to information and those who respect constraints while being relatively closed to information—are, perhaps, the more interesting leaders because they can at the same time benefit from and use the situation in which they find themselves. These are the strategic and pragmatic leaders. They are reflective of what the information processing literature has come to recognize as the cognitive manager (see, e.g., Suedfeld, 1992; Suedfeld, Tetlock, and Streufert, 1992; Suedfeld and Wallace, 1995). Cognitive managers engage in "conserving resources when spending them is unnecessary or futile, spending them when to do so leads to a net material or psychological gain" (Suedfeld, 1992:449). For these leaders, political timing is of the essence.

For the strategists who know what they want to do, information is sought concerning what the most feasible means are currently to reach that goal. For example, Hafez al-Assad of Syria was known to have three goals—to recover the Golan Heights, to guarantee the rights of the Palestinians, and to play a role in the region (Neumann, 1983–84; Pakradouni, 1983). But he wanted to achieve these goals while taking minimal risks. He "built his power stone by stone; he never rushes" (Dawisha, 1980:179). Thus, some have observed that the analyst could judge what issues were uppermost in Assad's mind by watching which foreign visitors came to Damascus. It was important to "size up the opposition," getting a sense of their positions and just how committed they were to their points of view before considering his next moves (Hermann, 1988b). The behavior of strategic leaders like Assad may seem unpredictable as they walk a fine line between actions that move them toward their goals while avoiding mistakes, failures, and disasters. As has been said of Assad, he took "care to hit the adversary without knocking him out and help the friend without really bailing him out, for the roles could be reversed one day" (Pakradouni, 1983:14).

For the pragmatic leader who respects the political constraints in the environment and seeks to work within them while at the same time having some idea about where he or she wants to take the government, the dilemma is to ensure that some progress is made toward a goal without stepping outside the bounds of one's position. If the time is right to push their own positions, they can do so; but such leaders can also accommodate to pressure if the time is not quite right. The observer may sense some indecisiveness as the pragmatic leader moves to uncover what will and will not work in a particular situation. Mohammad Khatami, the current president of Iran, is an illustration of a pragmatist (Hermann, 1999b). He ran for office on, and has championed throughout his tenure, a more moderate approach to Islamic law than currently governs the Iranian people, as well as its application in a fair and just fashion. But he also recognizes that the way the Iranian government is structured, he must share leadership with the Ayatollah Khamenei and the various clerical organizations that oversee the adjudication of laws and the selection of candidates for office. A cleric himself, Khatami has been trying to work within the system to ensure change can occur without overturning the Islamic revolution. He is searching for a way to reach his goal and maintain the support of his large, young population of supporters while not alienating the powerful clerics.

## Effects of Leadership Style

When we add the leader's motivation for action to his or her reactions to constraints and openness to information, as indicated in the last two columns of Table 1, we further differentiate leadership styles by denoting what individuals will do who are more concerned with the issues facing their governments vs. what individuals will do who are more interested in the responses of relevant constituencies and audiences. Although certain leaders have the facility to move between these two orientations, most feel more comfortable emphasizing one or the other (see Hermann, 1999a). In interaction with knowledge about reactions to political constraints and openness to information, these orientations suggest what lies at the heart of a leader's political agenda. Thus, crusaders who have a cause or problem to solve are likely to focus on expanding their span of control over resources and/or geographical space—empire, sphere of influence, and hegemony are important parts of their worldview—in order to increase their ability to gain future leverage in a particular domain. Crusaders who crave relationships and influence over others seek to convert others to their position or point of view—the more converts the greater the feeling of success. These expansionists and evangelists have little use for those who cannot understand the urgency of their concerns; they identify with their goals completely, at times becoming isomorphic with the positions of their countries and willing to risk their offices for what they believe is right and just. Their positions should prevail because they know what is best for all concerned. Those who cross such leaders are considered the equivalent of traitors. The expansionists and evangelists are not very concerned about the political environment around them except as it impedes their progress toward their goals. Their behavior is relatively predictable and consistent over time.

This last statement is not applicable to leaders with the other leadership styles described in Table 1. The current situation and state of the political context play a bigger role for them. The leadership styles are suggestive of what becomes important to the leader to assess in the situation and where the analyst may want to look to understand what is happening. Thus, leaders with an incremental leadership style are interested in maintaining control over what they do in foreign policy and having the flexibility and maneuverability to move slowly or quickly depending on the circumstances to increase the probability that they can achieve what they want. They are interested in any action, however modest, that will inch them toward their goals as long as said activity does not restrict their movement in the future. Leaders with a charismatic leadership style accept that perceptions of power and authority are often in the "eye of the beholder" and are desirous of ensuring that important constituencies and institutions understand and support what they are doing before, and even as, they are engaged in particular foreign policy activities. These leaders look for ways to enhance or, at the very least, to maintain their image in the eyes of certain constituencies. They are not averse to using diversionary tactics to consolidate their support and approval ratings. Both these leadership styles promote strategic and deliberative behavior; the particular setting and circumstances shape how these leaders will work to reach their goals. They know what they want to do; at issue is whether or not the current context indicates such behavior is feasible and likely to be successful.

Leaders with directive and consultative leadership styles have a political agenda but believe they must work within the domestic and international constraints that shape their office and their government's position in the international arena. They must pragmatically deal with the parameters that define their political space. Those with a directive leadership style focus on guiding policy deliberations in a direction that is responsive to their goals and what they perceive are important issues for the country to consider and address. Such leaders appear to intuitively understand, however, that there are bounds on their actions and are intent on respecting such limits while still moving to deal with current problems. The challenge becomes how to take the initiative or respond quickly and decisively when rules and procedures define what is possible and are intended to slow down the decision-making process. For leaders with a consultative leadership style, the people who have the potential of blocking or making action more difficult become the focus of attention, not the issue or topic under discussion. It is important for these leaders to become the hub of any information network so that they can monitor who supports or is in opposition to what they think should be done. Calculations are made about engaging in specific activities based on the extent of support and opposition among those to whom one is beholden for one's position.

As we noted previously, leaders who both respect constraints and are open to information are the most affected by the context and cues in the current situation in deciding what to do in foreign policy. They are the most buffeted by the political winds. And they tend to exhibit a reactive or an accommodative leadership style. Leaders with a reactive leadership style respond to how they view the particular problem they are facing can be managed given the current resources and political support that they have. These leaders attempt to be rational as they try to maximize what is possible while minimizing any costs to themselves and their chances of remaining in office. Problems are dealt with on a case-by-case basis; planning is considered difficult because one cannot take into account all possible permutations of events. It is the event not considered that will come to pass. For leaders with an accommodative leadership style, consensus-building and finding some compromise are the most relevant political skills. At issue is who are the relevant constituencies with regard to the current problem; how accountable is the leader to them? What actions will co-align the needs and interests of these important others? Is there a position these particular constituencies could support and around which they could rally? Others' positions and views become important in shaping what is done as the accommodative leader strives to build a consensus that will be acceptable.

## *Two Caveats*

The discussion to this point has described leadership styles that are derived from extreme scores on the three variables: reaction to constraints, openness to information, and motivation for action. Since each of these variables represents a possible continuum, the leadership styles in Table 1 should be considered ideal types. In considering what leadership style best characterizes a particular predominant leader, the analyst should select the one closest to the variables on which the person appears high or low. Where the individual seems more moderate, it is feasible to assume that he or she could move between the leadership styles for that variable.[3] Take as an example Slobodan Milosevic. He certainly challenged the political constraints in his environment and appeared to have a perspective through which he viewed the world, yet for quite a while he moved fairly easily between expanding his own power and control and enlivening the Serbs' sense of nationalism and preeminence (Hermann, 1999c). He manifested both expansionistic and evangelistic leadership styles, using one style to bolster the goal of the other.

One further caveat is important. The leadership styles in Table 1 can be applied to domestic as well as foreign policy. Two of the authors (Hermann, 1980, 1984, 1988a; Hermann and Preston, 1994; Preston, 2001) have discovered that leaders' styles, however, may change across these two domains depending on their degree of expertise in each. Whereas leaders may challenge constraints in the domain in which they have experience, the opposite may hold where they have little experience. In this arena, they are, in effect, learning on the job and may be more cautious and feel more constrained. And whereas leaders may be more open to listen and take advice when they have little experience, they may believe they know what needs to be done with experience. Leaders can be much more reliant on situational cues and their advisers when they are inexperienced than as they gain expertise. Consider how much more

---

[3] By using the assessment-at-a-distance technique mentioned in footnote 2, researchers can determine numeric scores for a leader on the three variables in Table 1 and compare that leader's scores with those of other leaders in the region or culture as well as with a composite set of scores for 150 national leaders. These assessments can be further contextualized by examining how a leader's scores may differ when talking before different audiences, being interviewed domestically or internationally, and discussing different topics, as well as in settings that vary as to the degree of spontaneity they afford the speaker. Instructions regarding how to use this technique are available in Hermann (1999a).

comfortable George H. W. Bush was in exercising his authority and control in dealing with foreign policy problems than he was in domestic politics; most of his positions prior to becoming President of the United States dealt with foreign policy (e.g., UN Ambassador, Director of the Central Intelligence Agency, chief of the U.S. Liaison Office after the opening to China) (Preston, 2001). As Bush observed after negating the advice of many following the uprising in Tiananmen Square, "I know China. . . . I know how to deal with them" (Duffy and Goodgame, 1992:182). After all, he, not them, had had experience in dealing with the Chinese leadership! Such was not usually the case for Bush with regard to domestic policy.

## ILLUSTRATIONS OF THE PREDOMINANT LEADER DECISION UNIT

In our previous discussion we described the conditions under which a single, powerful individual can become a predominant leader as well as proposed the importance of a leader's sensitivity to the political context in discerning how he or she is likely to act when a predominant leader and considered eight different leadership styles that are related to variations in sensitivity. What happens when we apply this framework to some cases? Does its application help us understand what occurred in a particular situation and why? In what follows, we are going to examine four cases where we believe the decision unit was a predominant leader: (1) the recognition in 1975 of the Popular Movement for the Liberation of Angola (MPLA) by the Nigerian reformist military regime of General Murtala Mohammed; (2) the decision by the Egyptian cabinet in April 1973 to go to war against Israel; (3) the1965 decision by the Johnson administration to escalate United States involvement in Vietnam; and (4) the Bush administration's decisions regarding how to deal with the Iraqi invasion of Kuwait in August 1990. These cases illustrate the various conditions under which powerful leaders can become predominant as well as provide examples of a crusader, a strategist, a pragmatist, and an opportunist and four distinct leadership styles.

### *Nigerian Recognition of the MPLA* [4]

***Occasion for decision.*** In July 1975 Murtala Mohammed gained control of the Nigerian government in a bloodless coup against General Yakubu Gowon's regime. The palace coup took place in response to popular discontent about the relative anarchy that had come to characterize Nigerian society and policy-

---

[4] This section builds on a case study developed by Shaw with the assistance of John Inegbedion.

making. The regime change ushered in a new era in Nigeria as the two men differed markedly in their personalities and leadership style. Gowon, a Christian from a small tribe in the Plateau State, was described as a "patient man of gentle nature"; Mohammed, a devout Muslim from an aristocratic family in Kano, was characterized as "tough, inflexible, [and] strong-minded" (Aluko, 1981:242). While Gowon had pursued a pro-Western foreign policy, Mohammed called for an independent foreign policy, for a Nigeria that "took hard stands" on sensitive issues (Shepard, 1991:87).

Murtala Mohammed's opportunity to exert a strong, independent foreign policy came early in his tenure and grew out of the Angolan civil war and the untimely departure of the Portuguese colonial administration in November 1975. Portugal withdrew from Angola, disregarding the role it had agreed to play in facilitating the development of a government of national unity among the three liberation movements—MPLA, UNITA, and FNLA—seeking to control Angola after the colonizers left. In the ensuing chaos, the MPLA which had gained control of the capital—Luanda—declared Angola independent and formed a government. With the two superpowers in the Cold War supporting different liberation movements, reactions to the MPLA declaration were swift. The South African government, with the knowledge of their counterparts in Zambia and Zaire and the help of the United States, moved its army from its base in southern Angola toward Luanda with the objective of wresting control of the country from the MPLA which they viewed as Marxist and knew had Cuban backing and Soviet advisers.

Mohammed and members of his government viewed this move as a replay of South Africa's interference on the side of Biafra in the attempt to break up Nigeria during its civil war. When an appeal to Kissinger urging that the United States pressure South Africa to stop its advance went unheeded, the Mohammed regime found itself faced with a foreign policy problem and the need to make a decision. Was a government of national unity still possible and, if not, should Nigeria recognize the MPLA and fight to consolidate Angola's sovereignty by urging African countries and the international community to accord the MPLA diplomatic recognition? As Akinyemi (1979:155) observed, "if the new regime was hoping for a methodical and gracious transition from a leisurely and somewhat conservative foreign policy to a dynamic one ... the Angolan crisis came as a rude reminder that foreign policy crises are no respecter of domestic political pace."

***Decision unit.*** Faced with this occasion for decision, did Mohammed act as a predominant leader in this case? In other words, did he have the authority to commit the resources of the government without having his position reversed and did he exercise that authority in this instance? Given the internal and external groups and coalitions that have often presumed themselves to have the ability to act for the Nigerian government and the relative anarchy that has often characterized the country's regimes (see Shaw, 1987), this question becomes an appropriate one to ask. This "pluralism" was an issue for Mohammed as well; he had made his predominance a condition for accepting the coup makers' offer to lead Nigeria earlier that year. As he clearly put it to the junta, "if you are inviting me to be head of state, I'm not going to allow you to tie my hands behind my back [by consensus decision making]. I must have executive authority and run the country as I see best" (Garba, 1987:xiv). Because the group believed that only Mohammed could keep the country, particularly the armed forces, together, the junta acquiesced. He moved quickly to restructure the foreign policymaking bodies within the government to bring them under his control. Indeed, his role in foreign policymaking was so pervasive that "it was widely believed that Murtala unilaterally took the decision to recognize the MPLA government in November 1975" (Aluko, 1981:247).

In addition to structuring the regime with himself at the pinnacle of power, two other conditions point to Mohammed's being a predominant leader in this decision. The first is his general interest in Nigerian foreign policy. He came to office with the view that "Nigeria must be visible in the world"; his foreign minister was instructed to spend "one week out of every two abroad" (Garba, 1987:9). As one of the commanding officers who had brought the Nigerian civil war to a decisive end, Mohammed was reported to have had a substantial interest in defense policy. Indeed, he was a major proponent of an African High Command—a permanent force with the purpose of forestalling extra-continental intervention in African political conflicts (Inegbedion, 1991). The second condition is that his regime was faced with a crisis that posed both a threat and an opportunity for him. The South African intervention into the Angolan civil war was an especially significant problem for Mohammed since this move contradicted one of his foreign policy priorities: the elimination of colonialism and racism in Africa. If he was to demonstrate his desire for an activist foreign policy and compete for continental leadership, he needed to do something dramatic and quick. By pushing for an African solution to the Angolan problem, Mohammed could demonstrate that Nigeria had "assumed the mantle of continental leadership relinquished by Ghana twenty years before" (Kirk-Greene and Rimmer, 1981:14). It was important that he not delegate this responsibility but that he make the decision.

*Leadership style.* Having determined that Mohammed acted as a predominant leader in response to this occasion for decision, does his leadership style help us understand the decision he made? In other words, by ascertaining how he reacts to political constraints, how open he is to incoming information, and what motivates him to act, can we suggest what he is likely to do in this kind of situation?

In both the domestic and foreign policy arenas, Mohammed appears to have been a constraint challenger. As we have already noted, in assuming the position of head of state, he indicated he would not be constrained. To ensure his control over foreign policy, he centralized decision making within the Supreme Military Council under his direct supervision and disregarded the recommendations of the once powerful Ministry of External Affairs. In contrast to his predecessor's use of the Organization of African Unity (OAU) to build a multilateral consensus regarding international issues, Mohammed saw it as an instrument of Nigerian foreign policy (Aluko, 1981). In addition, he felt that "Nigeria should be more vociferous on South Africa and Third World issues, disregard regional or continental institutions, identify with the Third World, oppose the global establishment, and challenge the West" (Shepard, 1991:87). He was intent on inculcating this new direction into Nigerian foreign policy.

Mohammed has been described as "bold, decisive, and pan-Africanist" (Agbabiaka, 1986:334). He came to office with a particular set of goals and viewed the political landscape through the lens of what he wanted to do. The South African military presence in Angola constituted not only a threat to the newly won independence of Angola but a serious breach of Nigeria's national defense. If Pretoria were allowed to occupy Angola directly or even indirectly through UNITA or FNLA surrogates, "it would only be a question of time before the adjoining states were gobbled up, and a direct threat posed to Nigeria" (Ogunsanwo, 1980:23). Indeed, P. W. Botha, South Africa's defense minister, had boasted that when they reached Luanda there would be little to prevent them from going on to Lagos. The only information Mohammed wanted was some indication of what needed to be done to build the necessary support to ensure general recognition of the MPLA. Information was used to facilitate implementation of a decision, not in the formation of the decision. In this regard, Mohammed was relatively closed to incoming information, particularly any that challenged his right to provide pan-African leadership. He was idealistic and, as such, knew what he wanted to do.

Mohammed came into office with what he perceived was a task to do: to develop an activist foreign policy and to have Nigeria assume its rightful position of leadership on the African continent. He was eager to move toward achieving these goals and was motivated to take whatever action would indicate his interest in tackling this task. He was looking for ways to indicate to other African governments that Nigeria was once more a player in continental affairs. Indeed, since coming to office, he had appointed one of the country's foremost analysts of Nigerian foreign policy, Bolaji Akinyemi, to formulate new guidelines for the country's external relations and commissioned a blue-ribbon committee of academics, commentators, and military officers to advise on the reform of the foreign policy system (Akinyemi, 1979; Garba, 1987).

As this discussion suggests, Mohammed was willing to challenge constraints, he had a set of goals that determined the kinds of information he sought in the environment, and he was motivated to tackle the task of creating an activist foreign policy. According to Table 1, Mohammed should evidence an expansionistic leadership style and act like a crusader. In this situation, his leadership style indicates that he should have been interested in trying to expand his and Nigeria's power and influence—to turn the crisis to the Nigerian government's advantage. Such a style should lead Mohammed to make a quick decision, seek loyal lieutenants to execute the decision, select a bold and dramatic action, and engage in what might be viewed as a risky maneuver.

*Foreign policy decision.* An examination of the Nigerian government's response to this occasion for decision indicates that once Mohammed was appraised of South Africa's invasion into Angola after the MPLA's declaration of independence, he believed that Nigeria must recognize the MPLA immediately. It was only in deference to diplomatic protocol and the felt need to appraise the American ambassador of the decision that Mohammed agreed to delay announcing his decision for twenty-four hours (Garba, 1987). With the announcement of Nigeria's recognition of the MPLA as the rightful government of Angola came a firm commitment of aid to the new government. "Once we accorded recognition, things moved with what came to be thought of as Murtala-esque speed . . . and anyone, particularly in the foreign ministry . . . who asked about a quid pro quo for Nigeria's staunch support was decisively overridden" (Garba, 1987:23). "We promised to give [the MPLA] everything from C-130 aircraft to fresh meat, and even took on the Gulf Oil Company on their behalf" (Garba, 1987:31). Moreover, all of Nigeria's instruments of statecraft were concentrated on implementing the decision. The Nigerian diplomatic corps and senior military officers were assigned the task of ensuring that Angola under the MPLA got the political recognition that comes with independence. "Nigeria's recognition of the MPLA was a key factor in the African collective swing to the MPLA and the eventual recognition of that party as the government of Angola" (Shaw and Aluko, 1983:174). By his decision, Mohammed had set the tone for an activist foreign policy which later saw the Nigerian government threaten to withdraw its Olympic team from the 1976 Montreal games and reject Anglo-American proposals for settling the constitutional deadlock in Rhodesia (Aluko, 1981). He had acted as a predominant leader and made a swift and dramatic decision reflective of a person with an expansionistic leadership style.

## Egyptian Decision to Attack Israel[5]

*Occasion for decision.* Middle East analysts agree that Egypt's war against Israel in October 1973 represents a major turning point in Arab-Israeli rela-

---

[5] This section builds on a case study done by Korany (1990).

tions. Not only did this war, which began on an Israeli religious holy day, Yom Kippur, have a national impact, it had wide-ranging regional and international ramifications. Many believe that the Arab-Israeli peace process and Sadat's dramatic visit to Jerusalem would not have happened without this war. Moreover, the decision produced an oil crisis as well as stagflation in the international system, and increased the role of the U.S. in the region, at the same time decreasing the influence of the Soviet Union.

Although there were a number of Egyptian decisions during the course of the October War, we are going to concentrate here on the decision to go to war. That is, we are interested in understanding the decision the Egyptian government made to go with a military solution to the stalemated situation it found itself in vis-à-vis Israel in 1973 rather than to continue the search for some diplomatic breakthrough. The October War poses something of a paradox because Egypt's president at the time, Anwar Sadat, is perceived as the peacemaker with Israel. And, indeed, he did embark on a visit to Jerusalem when Israel and Egypt were technically still at war, he was excluded from the Arab League for establishing a formal peace with Israel, and he probably paid—at least partly—with his life for his bold action. An argument can be made that Sadat was interested all along in a diplomatic solution to Egypt's problems with Israel but because of both domestic and international pressures could no longer ignore the war option and a military confrontation.

Egypt's defeat in the Six-Day War in 1967 compounded by the impasse in finding a political solution to its aftermath laid the foundations for the round of violence in October 1973. To the Egyptian leadership, the Six-Day War was a debacle both militarily and economically (see Korany, 1986a). Indeed, by the 1970s economic problems were beginning to constrain what the Egyptian government could do in foreign policy (Dessouki, 1991). With debt increasing by a yearly average of 28 percent, Egyptian foreign policy became focused on finding external help in paying for it. "In Egypt, ideological and political considerations were overshadowed by more immediate economic concerns" (Dessouki, 1991:161). Moreover, as a reaction to the growing economic strains, the public became more restless and vocal; demonstrations among both the military and students increased in the fall of 1972 as impatience grew with the fact that there was neither peace nor war with Israel and, as a result, their lives and prospects were grim. Among Egyptian officials there was a feeling of being under siege (see Rubinstein, 1977).

The Egyptian government under Sadat's leadership tried a number of different strategies to maintain their bargaining power and to attempt to find a solution to their economic problems that seemed tied up with the impasse with Israel. Much of the activity focused on restructuring Egyptian foreign policy away from the Soviet Union and toward rapprochement with the United States. These moves included a proposal in 1971 to reopen the Suez Canal and an expressed willingness to sign a peace treaty with Israel, expulsion of Soviet military advisers in the summer of 1972, and high-level talks with U.S. officials in the winter of 1973. All these efforts failed to produce meaningful results (see Quandt, 1977; Freedman, 1982; Dessouki, 1991). Sadat became frustrated with the ineffectiveness of his diplomatic initiatives to the West and convinced that as long as Egypt was perceived as a defeated party and Israel was in a position of superiority, the United States would do nothing.

Thus, in the spring of 1973, Sadat believed a decision needed to be made between diplomatic and military options. And events were pushing him toward a military solution. As Sadat remarked in a *Newsweek* interview (April 9, 1973), "the time has come for a shock. . . . Everything in this country is now being mobilized in earnest for the resumption of the battle—which is inevitable. . . . One has to fight in order to be able to talk." The Egyptian government and its leader, Anwar Sadat, were faced with an occasion for decision.

***Decision unit.*** Was Sadat a predominant leader in this case? Did he have the authority to commit the resources of the government without having his position reversed and did he exercise that authority in this instance? The Egyptian government is both presidential and the result of a military takeover. Constitutionally, and in practice, the presidency is the center of foreign policymaking and Egypt's four presidents (from General Naguib to Mubarak) are ex-army men. This latter fact has usually given excessive influence to the military in Egypt's decision making. Indeed, the thesis could be defended that Egypt's 1967 debacle was in great measure the result of the dispersion and rivalry between Nasser's presidential apparatus and a set of military fiefdoms. The rout and resultant humiliation of the army were the occasion for the resumption of authority by the president. And by all accounts the Egyptian armed forces of the 1970s were quite different from those of 1967; they were better educated, more professional, and trained in conditions as close as possible to the expected war environment (Heikal, 1975).

But because Sadat lacked Nasser's credentials and experience when he assumed the presidency in the fall of 1970, early in his presidency he was quite wary of the military. He was bent on curbing its political influence and maintaining it as a purely fighting force. Thus, during the three-year period between his arrival in power and the launching of the October War, Sadat changed the minister of defense three times before he found a person who was "professional, honest, [and] wholly above politics" (Heikal, 1975:184; see also Shazly, 1980). Moreover, he weathered an attempted coup and countercoup, ending up arresting prominent fellow leaders including the vice-president and placing them on trial for treason. Sadat was finally able to consolidate his authority in March 1973 when he formed a new cabinet with himself as prime minister as well as president. In taking both positions he could ensure that he had control over the policymaking apparatus when it came time to make a choice concerning how to deal with Israel (see Rubinstein, 1977; Freedman, 1982). Sadat viewed what happened between Egypt and Israel as having potential repercussions for both Egyptian domestic and foreign policy as well as making it easier or harder for him to retain power. The need to make a decision between diplomatic and military options in dealing with Israel was both a critical decision for the Egyptian government and one on which Sadat perceived his fate rested. He was not about to delegate authority to others when it came to making the decision. As Dessouki (1991:169) has observed, by early 1973 Sadat had the power to engage in "a highly personalized diplomacy . . . characterized by the ability to respond quickly and to adopt nontraditional behavior."

***Leadership style.*** Given that Sadat acted as a predominant leader in response to this particular occasion for decision, does knowledge about his leadership style aid us in understanding the decision he made? By determining how he reacts to political constraints, how open he is to incoming information, and what motivates him to act, can we propose what he is likely to urge on his government?

An assessment-at-a-distance of Sadat's leadership style (see Snare, 1992 for details) indicates he was likely to challenge the political constraints he perceived in his environment but was interested in doing so more behind the scenes than directly. Only when such activity was not having the desired effect would Sadat move to take a bold action (e.g, the expulsion of the Soviet advisers in the summer of 1972). Sadat "displayed an adeptness at balancing and reconciling political rivals" and a "sense of timing"; he worked to coax others to go along and to forge a consensus where such was feasible (Rubinstein, 1977:217, 238). He was prepared to exercise what he viewed as Egypt's leadership position in the Arab world—"a property that [he perceived] could not be challenged or taken away" (Dessouki, 1991:167)—to restore the territories occupied by Israel in 1967 and to deal with his country's dire economic problems.

The data in the leadership style assessment-at-a-distance profile also suggest that Sadat was sensitive to both confirmatory and disconfirmatory information in his political environment. He perceived himself to be balancing a number of domestic and foreign policy demands, trying to co-align the various forces into a "workable" policy. Once, however, Sadat had convinced himself of what would work, he expected "concrete solutions to flow automatically from political level agreement on the essentials" (Vance, 1983:174). He knew where he wanted to go in broad outlines; the detail and timing grew out of the particular context of the moment.

In his motivation for action, the assessment-at-a-distance data denote a focus on maintenance and survival of his country. His policymaking was intended to ensure that Egypt could survive economically and militarily. A number of scholars talk about Sadat's courtship of the West, the Soviets, and the Arab world as he sought to find a way to deal with the war of attrition facing his country in 1972 (e.g., Quandt, 1977; Rubinstein, 1977; Dessouki, 1991). He perceived that he needed the support of these others to be able to tackle Egypt's problems. Sadat's general affableness and desire for approval as well as his enjoyment of crowds and the spotlight lend support to the importance of relationships in both his political and personal life.

As this discussion suggests, Sadat was willing to challenge constraints but was open to information from his environment regarding what was possible and how far he could push at any point in time. Moreover, he was interested in building and maintaining relationships with the appropriate people and entities he believed could ease his domestic and foreign policy problems. Given this profile, according to Table 1 Sadat should exhibit a charismatic leadership style. As a predominant leader with the choice between engaging in more diplomacy or going to war in early 1973, the framework would expect Sadat to act strategically and, while making a general decision for war, to consider how to enhance the chances of success by including others in the process and the activity. Having made the decision he would choose that moment to implement it when he believed he had the relevant others onboard ready to participate and, in turn, enhance the likelihood for success.

*Foreign policy decision.* Accounts of Sadat's policymaking during the buildup to the Yom Kippur War indicate that the decision to go to war was made and ratified by the cabinet during April 1973. There was a sense at the time that a military confrontation with Israel was no longer a moral necessity but a political one. But the decision was not implemented immediately because Sadat perceived that he needed to prepare the political terrain first. He "embarked on an ambitious policy of enormous complexity. The intricacy of the design was only dimly perceived at the time" (Rubinstein, 1977:217). His strategy was intended to ensure that his own people and military were ready for what was going to take place, the flow of Soviet arms was adequate to the task, he had the economic and political support of the oil-rich Arab states and their willingness to use the oil card if necessary, he had secured an alliance with Syria that enabled a surprise attack on two fronts simultaneously, and enough diplomatic activity was in place to keep the United States and Israelis off guard as to Egypt's plans. Sadat considered any war to be limited in scope; he was intent on doing what it took to improve the negotiating odds for Egypt with Israel and the United States. That he went a long way toward achieving his goal with all his maneuvering after making the decision to engage in a military confrontation with Israel is evident in the following observations: "The prevailing attitude toward the Arab world held by [American] policymakers was challenged by the October war" (Quandt, 1977:201); indeed, "it required the October war to change United States policy and to engage Nixon and Kissinger in the search for an

Arab-Israeli settlement" (Quandt, 1977:164). "The Arabs regained their dignity and no longer feared to negotiate as an inferior, defeated party" (Safran, 1989:390).

## *Escalation of U.S. Involvement in Vietnam* [6]

***Occasion for decision.*** One of the most studied and hotly debated foreign policy decisions of Lyndon Johnson's presidency is the decision in July 1965 to dramatically escalate American troop involvement in Vietnam (e.g., Thomson, 1968; Hoopes, 1969; Janis, 1972; Kearns, 1976; Berman, 1982, 1989; Burke and Greenstein, 1991). Indeed, some have argued that the escalation was a critical juncture in the Vietnam War and in the Johnson presidency. In approving General Westmoreland's request for forty-four battalions of ground troops (over 125 thousand men) for use in South Vietnam to halt the Viet Cong offensive and "restore the military balance" vis-à-vis Communist North Vietnam, Johnson became politically trapped in a continually escalating spiral of involvement in a war he did not want. At the same time, his Great Society domestic programs which represented his true policy interests were left largely unimplemented and drained of resources by the conflict in Indochina (see Johnson, 1971; McPherson, 1972; Kearns, 1976). Instead of leaving the legacy in domestic policy that he had intended, the Johnson presidency is more often defined, by historians and the public, by his connection to the Vietnam War.

The debate over increasing the number and role of U.S. ground troops in Vietnam in the summer of 1965 was a significant phase in the "Americanization" of the war. The immediate problem confronting the Johnson administration was the deterioration in the situation in South Vietnam. By June 1965, the failure of the U.S. air campaign against North Vietnam had become apparent to the White House. Instead of decreasing North Vietnam's resolve and determination, the bombing was having just the opposite effect (Berman, 1982). The unstable South Vietnamese government had changed once more. The Viet Cong had executed an American prisoner of war and bombed a riverboat restaurant near the American embassy in Saigon. Dire predictions were being made for Vietnam by Ambassador Taylor and General Westmoreland unless there was a significant increase in American forces in the area. The new South Vietnamese leadership echoed the call for additional troops.

> The problem now facing President Johnson was different from that which had faced his predecessors. Within a matter of weeks South Vietnam would fall to the Communists without a substantial ground commitment by the United States.

---

[6] This section draws on materials from Ripley and Kaarbo (1992) and Preston (2000, 2001).

Was the United States committed to saving South Vietnam, preventing a Communist takeover, or saving face? (Berman, 1982:77)

An occasion for decision was at hand. By early July Johnson realized he and his advisers had to reach a final decision regarding whether to escalate or reduce the U.S. commitment to the war. And, if escalation were chosen, to decide the size of the deployment, if reserves would be called up, and what the ceiling would be to the overall American commitment.

***Decision unit.*** Faced with this occasion for decision, did Johnson act as a predominant leader in this case? A hallmark of Johnson's style, both in the Senate and oval office, was a tightly hierarchical, centrally controlled organizational structure among his advisers that facilitated his maintaining dominance over the policy environment (McPherson, 1972; Kearns, 1976; McNamara, 1995; Preston, 2000). Johnson insisted on being his own chief of staff and at the center of all lines of communication. He structured the nature of debates among his advisers and made all the final decisions himself. Nothing occurred in the Johnson White House without the president's approval. No significant decisions were taken or policy initiatives adopted without his involvement. As Dean Rusk (1969:38–39) observed, "as far as Vietnam is concerned, President Johnson was his own desk officer . . . every detail of the Vietnam matter was a matter of information to the President, and the decisions on Vietnam were taken by the President." Moreover, after the Gulf of Tonkin Resolution of 1964, Johnson had virtually a blank check to take nearly any military action he desired in Vietnam due to the vague nature of the resolution (see Berman, 1982; Ambrose, 1993). Although Johnson did delegate responsibility for gathering information and formulating options to his advisers with regard to Vietnam, there is no evidence that he ceded any authority for committing the resources of the government. His advisers were there to propose and evaluate options, not to choose (Goldman, 1969).

***Leadership style.*** Having determined that Johnson acted as a predominant leader in response to this occasion for decision, does information on his leadership style help us understand the decision he made? In other words, if we ascertain how he reacts to political constraints, how open he is to incoming information, and what motivates him to act, can we say anything about the nature of what he will propose to do?

Although Johnson's desire to be in control of his policy environment might at first glance suggest that he would challenge constraints, it is more accurate to conceive of him as working within the constraints he perceived were operative in any political setting. Johnson possessed a subtle appreciation of power in Washington and recognized the need to be attentive to Congress while at the same time taking into consideration the domestic political scene. He preferred not to be the first person to stick his neck out on a tough issue. Instead, Johnson tended to move only after the waters had been tested and others had staked out the advance positions. "He entered into the labors of others and brought it about" (McPherson, 1995). As McGeorge Bundy (1993) noted, Johnson was always the legislator gathering information to facilitate building a majority and negotiating his way toward a decision that would not alienate those whose support he needed. Even the so-called Johnson treatment was intended, in his own words, to "let me shape my legislative program to fit their needs and mine" (Kearns, 1976:186). There is some evidence that his lack of experience in foreign affairs led him to be more deferential toward the constraints in the foreign policy environment and less forceful in the exposition of his own views than he was in considerations of domestic policy. In the foreign policy arena, "he was insecure, fearful, his touch unsure. . . . [H]e could not readily apply the powerful instruments through which he was accustomed to achieve mastery" (Kearns, 1976:256).

Johnson possessed a largely undifferentiated image of the world. He relied heavily on stereotypes and analogies, processing most of his information about foreign affairs through relatively simple lenses. Thus, for example, Johnson tended to view Vietnam in straightforward ideological terms; it was a conflict that involved freedom vs. communism, appeasement vs. aggression (Kearns, 1976:257). Johnson's simplified worldview not only envisioned American values as having universal applicability abroad but held that these values were so clearly correct that there was a worldwide consensus regarding their positive nature. Moreover, he saw parallels between Vietnam in the summer of 1965 and Munich. He explained his inability to withdraw from Vietnam because history told him that "if I got out of Vietnam and let Ho Chi Minh run through the streets of Saigon, then I'd be doing what Chamberlain did in World War II" (Khong, 1992:181). And even though Johnson was noted to be a voracious consumer of information, he was very selective in the type of information he sought from his environment, focusing primarily on information that would assist him in passing or implementing a program (Bundy, 1993; McPherson, 1995).

Johnson has been described by colleagues as highly task oriented and strongly driven to accomplish his policy objectives, judging all his daily activities by one yardstick—whether or not they moved him toward his goals (Rusk, 1969; Califano, 1991). Nothing was more important than accomplishing what he wanted to accomplish. "He was in a hurry and wanted to see the results of something . . . really wanted to change the world, to be the best President ever" (Christian, 1993). Johnson's work habits were legendary among his colleagues who frequently noted the intensity with which he approached his job. As Rusk (1969:1–2) observed, "he was a severe task-master, in the first instance of himself." Johnson was a "doer." His emphasis on solving problems is best illustrated by his inevitable response to staffers who entered his office describing some terrible problem or chaotic situation without at the same time proposing a solution or course of action. Johnson would stare at them and proclaim: "Therefore!" As Christian (1993) noted, "they learned pretty quick that you better come in with an action to address it [the problem]. You couldn't just get away with saying something's wrong."

As this discussion suggests, Johnson worked within the political constraints he perceived defined his political environment, being caught up as were many policymakers of his era in the lesson from Munich not to stand back while smaller nations were absorbed by enemies who would never be satiated. He believed in the validity of containment as a policy and used it as a lens through which to interpret information about Vietnam and other countries in the international arena. And he was task oriented, always interested in making decisions that would deal with the perceived problem of the moment. Given this profile, according to Table 1, Johnson should exhibit a directive leadership style for this particular occasion for decision.[7] As a predominant leader decision unit, we would expect him to guide policy along paths consistent with his own views but to try to do so within the political parameters of the current situation. For leaders with this style, it is important to know something about the views of their advisers and the constituents they perceive are relevant to the problem at hand since these individuals will help to define the nature of the political context at any point in time. Given their pragmatic inclinations, leaders with a directive leadership style will take their cues about what is feasible and doable from the options, debate, and discussion around them. How extreme any foreign policy decision is likely to be will depend on the range of options under consideration and the potential for a satisfactory compromise or consensus among those in the political setting when a choice needs to be made.

*Foreign policy decision.* An examination of policymaking regarding the decision to send forty-four battalions of U.S. soldiers to Vietnam indicates that the decision to take this action was not made until July 27, though Johnson's inner circle was involved in debating the appropriate course of action from late June onwards. The president was ascertaining the nature of the political terrain during the intervening time period, testing out his own ideas as well as determining where others stood and with what degree of commitment. After a contentious meeting of the National Security Council (NSC) on June 23rd in which his advisers strongly disagreed over the next course of action in Vietnam, Johnson

---

[7] Assessment-at-a-distance data collected on Johnson supports this interpretation of his reaction to political constraints, the way he processes information, and his motivation for action (see Hermann, 1984; Preston, 1996, 2000).

requested the protagonists to draft separate proposals arguing for their particular position. In the course of a week he received three different proposals: one from Under Secretary of State George Ball arguing against deployment and making the case for "cutting our losses" and withdrawing U.S. troops from South Vietnam, the second from Secretary of Defense Robert McNamara that favored the deployment of 175 thousand American troops in 1965 and an undetermined additional number in 1966 as well as a significant call-up of the reserves (approximately 235 thousand reserves and national guards); and the third from Assistant Secretary of State William Bundy proposing a "middle way," or the deployment of 75-85 thousand troops and holding off further deployment decisions until the effects of the initial step could be ascertained over the summer while still continuing the existing bombing campaign.

At a meeting of the NSC on July 21, Johnson emphasized that he wanted a thorough discussion of all the options by his advisers "so that every man at this table understands fully the total picture."[8] The session was marked by a great deal of give and take. As the meeting progressed, it became clear that Ball's argument had failed to sway the others. As Bundy observed toward the conclusion of the session: "The difficulty in adopting it [Ball's option] now would be it is a radical switch without evidence that it should be done. It goes in the face of all we have said and done." Johnson held a series of meetings following this one in which he challenged the position of the Joint Chiefs of Staff for calling for an even larger escalation than that proposed by McNamara; talked with members of the NSC about strategies for approaching Congress, selling the policy to the public, and calling up the reserves; and reacted to Clark Clifford's arguments for getting out of Vietnam.

When he came to the meeting of the NSC on July 27th, Johnson was ready to make the final decision, having heard the debate over increasing involvement and the likely impact on domestic politics and public opinion. After observing that the situation in Vietnam continued to deteriorate, Johnson laid out what he believed to be the five choices the government had and remarked to his advisers that his preference was to give "the commanders the men they say they need" but to "neither brag about what we are doing or thunder at the Chinese Communists and the Russians" and to engage simultaneously on the diplomatic side in working to bring the Viet Cong and North Vietnamese to the negotiating table. He did not ask for the authority to call up the reserves because he perceived such an action would be unpopular with the public. With this decision,

---

[8] Meeting on Vietnam in Cabinet Room, 10:40 a.m., July 21, 1965 from "July 21–27, 1965 Meetings on Vietnam" folder, Papers of Lyndon B. Johnson, Meeting Notes File, Box 1, Johnson Library. The rest of the discussion about this decision builds on materials from this folder in Meeting Notes File, Box 1 at the Johnson Library.

Johnson tried to develop a policy consistent with his own views of the situation but that also fit within the parameters in which he found himself politically. It was heavily influenced by the views of the political terrain of those who were involved in the discussions.

## U.S. Decision to Intervene in the Gulf Crisis [9]

***Occasion for decision.*** After weeks of regional tensions between Iraq and Kuwait and several ill-fated diplomatic efforts to defuse the crisis by both United States and Arab leaders, Iraq invaded Kuwait on August 1, 1990, quickly overrunning the country. Although the U.S. intelligence and defense communities had observed Iraqi military movements in the days prior to the invasion and alerted policymakers, the Bush administration was still caught by surprise at the sudden Iraqi assault (Baker, 1995; Powell, 1995; Bush and Scowcroft, 1998). George H. W. Bush and his advisers were faced with an unanticipated problem and rushed to take some stop-gap measures such as calling for an emergency session of the U.N. Security Council, freezing all Iraqi and Kuwaiti assets in the United States, and ordering warships dispatched to the Gulf. They began deliberations the next day on possible U.S. policy responses to the Iraqi invasion. An occasion for decision had been forced upon them. All were concerned about the potential of Iraq moving into Saudi Arabia and the implications of the current crisis for world oil supplies. Brent Scowcroft (Bush and Scowcroft, 1998:315, 317) later recalled about this first meeting that it "was a bit chaotic" since they "really did not yet have a clear picture of what was happening on the ground" but he "was frankly appalled at the undertone of the discussion which suggested resignation to the invasion and even adaptation to a *fait accompli*." "[T]he discussion did not come to grips with the issues" (Powell, 1995:463). Bush's response was that "we just can't accept what's happened in Kuwait just because it's too hard to do anything about it" (Woodward, 1991:229). Thus began a series of formal and informal meetings as the president and his advisers wrestled with what to do.

***Decision unit.*** Faced with this occasion for decision, did Bush act as a predominant leader in this case? In other words, did he have the authority to commit the resources of the government without having his position reversed and did he exercise that authority in this instance? Because of his extensive foreign policy experience before becoming president, Bush had immense self-confidence in his abilities to deal with complex foreign policy issues and showed far greater interest in being actively involved in foreign than in domestic policymaking

---

[9] This section draws on material in Preston (2001); the assessment-at-a-distance data are described in detail in Winter et al. (1991).

(Rockman, 1991; Barilleaux, 1992; Crabb and Mulcahy, 1995; Hermann and Preston, 1999). Foreign policy was Bush's "meat" while most everything else was "small potatoes"; it was the role of "foreign-policymaker-in-chief" that most captured Bush's interests (Rockman, 1991:12). Bush took an active role in setting the overall foreign policy agenda, framing specific foreign policy issues, and shaping most final foreign policy decisions in his administration (Crabb and Mulcahy, 1995). Although he generally sought to obtain consensus among his advisers for his policy views, he was also comfortable relying solely upon his own policy judgments if these conflicted with the views of his experts (Barilleaux, 1992; Baker, 1995). During the decision making regarding the Iraqi invasion into Kuwait, Bush was observed to have played "his own Henry Kissinger" in terms of his personal involvement in the diplomacy required to maintain a coalition (Crabb and Mulcahy, 1995:254). He wanted to be in control and to know all the details regarding the situation (Woodward, 1991). In what was perceived as a crisis, he intended to be the predominant leader.

*Leadership style.* Given that Bush appears to have acted as a predominant leader in response to this occasion for decision, does learning about his leadership style help us understand what he decided to do? By determining how he reacts to political constraints, how open he is to incoming information, and what motivates him to act, can we make any proposals about what he is likely to urge the government to do?

An assessment-at-a-distance of Bush's leadership style (see Winter et al., 1991; Preston, 2001) indicates that he is responsive to the political constraints in his environment. He sought, where possible, to develop consensus among his advisers through a willingness to compromise on policy specifics, if not overall goals. He has been called "a low-key version of Lyndon Johnson" (Rockman, 1991:18), negotiating quietly outside the glare of publicity with as little rancor as possible and then announcing a compromise or consensus. Bush was noted to be extremely sensitive to the political arena when making decisions, very focused on maintaining the support of important constituents; he was "by instinct a retail politician that takes care not to alienate anyone" (*New York Times*, 1990:3). As a result, he placed great emphasis on gathering feedback from the external environment and in cultivating an extensive informal network of contacts. With such information, he could choose the "middle path" on issues and preserve good relations with important others (Woodward, 1991; Duffy and Goodgame, 1992). "Bush did not dream impossible dreams or commit himself to unattainable objectives" (Crabb and Mulcahy, 1995:256). He worked within the constraints of the political situation in which he found himself.

The assessment-at-a-distance data also suggest that Bush was open to information from the political environment. Indeed, he established an "honest broker"-style national security adviser to "objectively present to the president the views of the various cabinet officers across the spectrum" (Bush and Scowcroft, 1998:18) and an extensive network of individuals that included people both inside and outside the administration as well as foreign leaders. He used the networks as a sounding board. "The President called his principal allies and friends often, frequently not with any particular issue in mind but just to chat and exchange views on how things were going in general" (Bush and Scowcroft, 1998:61). During the Gulf crisis, Bush's personal diplomacy and policy discussions with regional leaders were extensive, broad, and detailed in scope. He believed that a "thousand shades of gray" existed in foreign affairs and was constantly probing his surroundings for data that permitted meaningfully interpreting the current political situation and climate before making a decision (Rockman, 1991; Crabb and Mulcahy, 1995).

As the previous discussion suggests, and the assessment-at-a-distance data indicate, building relationships was important for Bush. The observation that it is people, not ideas or issues, that drove Bush's interactions is a common one. He depended on personal relationships in building his formal and informal networks of contacts. Moreover, he was very cautious not to damage this personal network or upset important contacts. Indeed, he was interested in advisers who were team players and loyal to facilitate the building of consensus (Rockman, 1991; Woodward, 1991). Bush was widely viewed as a leader who emphasized the politics of harmony and conciliation through compromise rather than the politics of confrontation (Berman and Jentleson, 1991; Rockman, 1991; Barilleaux, 1992). He paid attention to maintaining relationships and relied on these ties to help him understand the political constraints in any situation and as sources of specific information about just what was happening at a particular point in time. He believed that personal diplomacy and leadership went hand in hand and that by developing personal relationships with constituents and other leaders one could gain cooperation, avoid misunderstandings, and obtain room to maneuver on difficult political issues.

In sum, Bush appears to have respected the parameters of the political environment in which he found himself, seeking to understand in full the nature of the particular occasion for decision that faced him by using his extensive network of contacts. His leadership style involved listening to his advisers, consulting by phone with relevant domestic and world leaders, and collecting as much information about the situation as possible. Building and maintaining relationships was critical to leadership. Given this profile, according to Table 1, Bush should exhibit an accommodative leadership style for this particular occasion for decision. As a predominant leader decision unit, we would expect him to focus on reconciling differences and building consensus among those involved in policymaking both within the country and outside, empowering these others to be part of the process, and, in turn, sharing accountability for what happens. Any decisions that are made are likely to represent a compromise among the options considered and are intended not to alienate or antagonize important constituencies. Of prime importance is ascertaining what is opportune in this particular political moment.

***Foreign policy decision.*** An examination of the U.S. policymaking process following the Iraqi invasion of Kuwait reveals that Bush and his advisers explored a range of options—international economic sanctions, a naval blockade of Iraq, deterrence of further Iraqi actions against Saudi Arabia, and a strike against Iraq—all the while seeking information. As Bush commented, "we didn't want to make statements committing us to anything until we understood the situation" (Bush and Scowcroft, 1998:317). He used his time to phone foreign as well as congressional leaders on the issue. From these contacts Bush was able to find out what particular individuals knew and where they stood; he also briefed them on other leaders' positions. In the process of moving these leaders toward action, he sought to build consensus, to listen to the concerns of the other leaders, and to avoid the appearance of dictating to them. Bush believed that "whatever we do, we've got to get the international community behind us" (Powell, 1995:464).

With the United Nations Security Council primed to pass Resolution 661 imposing economic sanctions on Iraq, Bush and his advisers turned their attention to military options in their meetings on August 3rd and 4th. There was much discussion and debate not only about what needed to be done militarily but also about the political consequences of various U.S. actions both domestically and internationally. As consensus grew that keeping Hussein from entering Saudi Arabia was an important objective of any U.S. action, considerations turned to the deployment of troops, how many, what kind, and where they would be deployed. After much give and take, with a number of his advisers indicating misgivings about some aspect of what was being proposed, Bush papered over the differences by approving both a naval blockade of Iraq and deployment of U.S. forces to Saudi Arabia contingent on King Fahd's approval (Bush and Scowcroft, 1998:328–329). With this action Bush had made the decision to intervene in Saudi Arabia to deter Iraq, although the ultimate shape of American policy and the fate of Kuwait had not yet been decided. And he began work on convincing the Saudi leadership to accept American troops on their soil.

## Conclusions

The four occasions for decision and leaders we have discussed here have indicated how leadership style can have an effect on what governments do in foreign policy when the decision unit is a single, powerful individual—a predominant leader. By learning how such leaders are likely to react to the political constraints they perceive in their environments, how open as opposed to selective they usually are in viewing incoming information, and whether they are more motivated to accomplish something or to build and maintain relationships, we gain the ability to ascertain leadership style and to determine how sensitive these people are likely to be to the political context.

An examination of the four cases suggests that differences in sensitivity can have a number of effects on decision making when the decision unit is a predominant leader.

- The decisions appear to become less defined and definitive as we move from crusaders like Mohammed to the opportunists like Bush and from the leaders with an expansionist style to those who are more accommodative. The goals are less visible in the decisions, and information about how the leader views his or her current political situation grows in importance.
- The people surrounding the predominant leader have a better chance of influencing the decision the more strategic, pragmatic, or opportunistic the leader is. Indeed, these types of leaders are dependent on cues from the environment in making their decisions. Such appears to be particularly the case if they value relationships. The issue becomes ensuring that all voices are being heard. What if there had been more policymakers of George Ball's persuasion in Johnson's advisory group; would they have swayed what Johnson thought or, at the least, suggested that there was a political constituency for that option? While the more strategic and opportunistic leaders are likely to seek out a range of opinions because of their openness to information, pragmatic leaders may get caught thinking they have consulted widely when their predilection for selective perception has limited whom and what they have heard.
- Political timing seemed the most important to the strategic leader, Sadat. He had a sense of what needed to happen before he could consider implementing his decision. The crusader here, Mohammed, was hell-bent to take action almost regardless of the consequences, while Johnson and Bush were working on developing the objectives they hoped to see occur even as they were responding to the events of the moment. The latter two knew more what they did not want to see happen than what they hoped would occur.
- Each of these occasions for decision was part of a sequence of such occurrences across a period of time. We have focused in these instances on one frame of a larger film or episode. We expect these four leaders with their various leadership styles to show differences in how they will react to setbacks or future successes. Mohammed as an expansionist will continue to press for his independent foreign policy in a variety of venues;

Sadat with his charismatic leadership style will evidence a foreign policy that "zigs and zags" as he tries different ways of resolving his country's economic woes by becoming tied to the West; Johnson being directive will work to keep U.S. foreign policy in Vietnam consistent with his image of containment policy as well as what he and those around him perceive are the relevant domestic and international pressures of the moment until there is an irreconcilable divergence among both domestic and international opinions on what to do; and Bush with his accommodative leadership style will be interested in making decisions that will maintain a broad consensus among his advisers and the formal and informal networks he has built at home and abroad—when that consensus begins to unravel and there is none that can be sustained in its place, the American intervention in the Gulf will be over.

The occasions for decision we have described were selected to illustrate how the predominant leader part of the decision units framework functions and to provide the reader with examples of four different types of leaders. The leaders were chosen because the authors had studied and written about them before. We have, however, merely laid the foundation for future research. It is important now to consider broadening the scope of our studies to include examining a larger set of leaders and linking assessment of their leadership styles to event data for governments for those situations where we would expect them to act as predominant leaders. Hermann and Hermann (1989) report a prototype of what such research might look like. The advent of software to assess both leadership style (Profiler+; see Young, 2000) and governments' actions (KEDS/TABARI; see Schrodt, Davis, and Weddle, 1994; Schrodt and Gerner, 2000) makes such studies more feasible. There is also merit in doing more intensive case studies where we explore individual leaders with the potential for predominance across a series of occasions for decision within the same problem domain as well as across a set of problems. Do the leaders always serve as the decision unit or only under certain conditions; does leadership style differ by domain, type of problem, degree of expertise; is there a consistency in the effects of leadership style on governments' behavior or do different types of feedback heighten or diminish a particular effect? Preston (2001) has begun such research on post-World War II American presidents but there is more to do. Does political structure and time in history affect the potency of the relationships proposed for this piece of the decision units framework? Given the lack of a monotonic relationship, noted earlier, between type of political structure and sensitivity to the political context, there should be leaders in democratic systems that manifest expansionist and evangelist leadership styles as well as leaders in autocratic systems that are reactive and accommodative. Are such leaders like their counterparts in the other type of political system? Hermann and Kegley (1995) have made some proposals about what we might expect to find that need exploration. As the reader will note, we have just begun to study the predominant leader decision unit. We welcome help in taking the next steps.

# REFERENCES

AGBABIAKA, TUNDE (1986) Murtala Remembered. *West Africa* (February):334.

AJZEN, I., AND M. FISHBEIN (1980) *Understanding Attitudes and Predicting Social Behavior.* Englewood Cliffs, NJ: Prentice-Hall.

AKINYEMI, A. BOLAJI (1979) "Mohammed/Obasanjo Foreign Policy." In *Nigerian Government and Politics Under Military Rule*, edited by Oye Oyediran. London: Macmillan.

ALEXANDER, ROBERT J. (1982) *Romulo Betancourt and the Transformation of Venezuela.* New Brunswick, NJ: Transaction Books.

ALUKO, OLAJIDE (1981) *Essays on Nigerian Foreign Policy.* London: Allen & Unwin.

AMBROSE, STEPHEN E. (1993) *Rise to Globalism: American Foreign Policy Since 1938*, 7th ed. New York: Penguin.

AXELROD, ROBERT (1976) *Structure of Decision: The Cognitive Maps of Political Elites.* Princeton, NJ: Princeton University Press.

BAKER, JAMES A. (1995) *The Politics of Diplomacy: Revolution, War, and Peace, 1989–1992.* New York: G. P. Putnam's Sons.

BARBER, JAMES DAVID (1965) *The Lawmakers: Recruitment and Adaptation to Legislative Life.* New Haven, CT: Yale University Press.

BARBER, JAMES DAVID (1977) *The Presidential Character: Predicting Performance in the White House.* Englewood Cliffs, NJ: Prentice-Hall.

BARILLEAUX, RYAN J. (1992) "George Bush and the Changing Context of Presidential Leadership." In *Leadership and the Bush Presidency: Prudence or Drift in an Era of Change?* edited by Ryan J. Barilleaux and M. W. Stuckey. New York: Praeger.

BASS, BERNARD M. (1981) *Stogdill's Handbook of Leadership.* New York: Free Press.

BASS, BERNARD M. (1985) *Leadership and Performance Beyond Expectations.* New York: Free Press.

BASS, BERNARD M. (1997) Does the Transactional-Transformational Leadership Paradigm Transcend Organizational and National Boundaries? *American Psychologist* **52**:130–139.

BASS, BERNARD M., AND E. R. VALENZI (1974) "Contingent Aspects of Effective Management Styles." In *Contingent Approaches to Leadership*, edited by J. G. Hunt and L. I. Larson. Carbondale: Southern Illinois University Press.

BENNIS, WARREN G., AND BURT NANUS (1985) *Leaders: The Strategies for Taking Charge*. New York: Harper and Row.

BERMAN, LARRY (1982) *Planning a Tragedy: The Americanization of the War in Vietnam*. New York: W. W. Norton.

BERMAN, LARRY (1989) *Lyndon Johnson's War: The Road to Stalemate in Vietnam*. New York: W. W. Norton.

BERMAN, LARRY, AND BRUCE W. JENTLESON (1991) "Bush and the Post-Cold-War World: New Challenges for American Leadership." In *The Bush Presidency First Appraisals*, edited by Colin Campbell and Bert A. Rockman. Chatham, NJ: Chatham House.

BROWNING, RUFUS P., AND HERBERT JACOB (1971) "The Interaction Between Politicians' Personalities and Attributes of Their Roles and Political Systems." In *A Source Book for the Study of Personality and Politics*, edited by Fred I. Greenstein and Michael Lerner. Chicago: Markham.

BUNDY, MCGEORGE (1993) Interview with Thomas Preston, November 18.

BURKE, JOHN P., AND FRED I. GREENSTEIN (1991) *How Presidents Test Reality: Decisions on Vietnam, 1954 and 1965*. New York: Russell Sage Foundation.

BURNS, JAMES MACGREGOR (1978) *Leadership*. New York: Harper and Row.

BUSH, GEORGE H. W., AND BRENT SCOWCROFT (1998) *A World Transformed*. New York: Alfred A. Knopf.

BYARS, R. S. (1973) Small-Group Theory and Shifting Styles of Political Leadership. *Comparative Political Studies* **5**:443–469.

BYMAN, DANIEL I., AND KENNETH M. POLLACK (2001) Let Us Now Praise Great Men: Bringing the Statesman Back In. *International Security* **25**:107–146.

CALIFANO, JOSEPH A., JR. (1991) *The Triumph and Tragedy of Lyndon Johnson: The White House Years*. New York: Simon and Schuster.

CAMPBELL, COLIN (1986) *Managing the Presidency: Carter, Reagan and the Search for Executive Harmony*. Pittsburgh, PA: University of Pittsburgh Press.

CHEMERS, M. M. (1997) *An Integrative Theory of Leadership*. Mahwah, NJ: Lawrence Erlbaum.

CHRISTIAN, GEORGE (1993) Interview with Thomas Preston, August 4.

CRABB, CECIL B., JR., AND KEVIN V. MULCAHY (1986) *Presidents and Foreign Policy Making: From FDR to Reagan*. Baton Rouge: Louisiana State University Press.

CRABB, CECIL B., JR., AND KEVIN V. MULCAHY (1995) George Bush's Management Style and Operation Desert Storm. *Presidential Studies Quarterly* **15**:251–265.

DAWISHA, AIDEED I. (1980) *Syria and the Lebanese Crisis*. London: Macmillan Press.

DESSOUKI, ALI E. HILLAL (1991) "The Primacy of Economics: The Foreign Policy of Egypt." In *The Foreign Policies of Arab States: The Challenge of Change*, 2nd ed., edited by Baghat Korany and Ali E. Hillal Dessouki. Boulder, CO: Westview Press.

DESTLER, I. M., HARUHIRO FUKUI, AND HIDEO SATO (1979) *The Textile Wrangle*. Ithaca, NY: Cornell University Press.

DiRENZO, GORDON J. (1977) "Politicians and Personality: A Cross-Cultural Perspective." In *A Psychological Examination of Political Leaders*, edited by Margaret G. Hermann. New York: Free Press.

DRIVER, MICHAEL J. (1977) "Individual Differences as Determinants of Aggression in the Inter-Nation Simulation." In *A Psychological Examination of Political Leaders*, edited by Margaret G. Hermann. New York: Free Press.

DUFFY, MICHAEL, AND DAN GOODGAME (1992) *Marching in Place: The Status Quo Presidency of George Bush*. New York: Simon and Schuster.

EVANS, PETER B., HAROLD K. JACOBSON, AND ROBERT D. PUTNAM (1993) *Double-Edged Diplomacy: International Bargaining and Domestic Politics*. Los Angeles: University of California Press.

FAZIO, RUSSELL H. (1986) "How Do Attitudes Guide Behavior?" In *Handbook of Motivation and Cognition: Foundations of Social Behavior*, edited by Richard Sorrentino and E. Tony Higgins. New York: Wiley.

FIEDLER, FRED E. (1967) *A Theory of Leadership Effectiveness*. New York: McGraw-Hill.

FIEDLER, FRED E. (1993) "The Leadership Situation and the Black Box in Contingency Theories." In *Leadership Theory and Research: Perspectives and Directions*, edited by M. M. Chemers and R. Ayman. New York: Academic Press.

FIEDLER, FRED E., AND M. M. CHEMERS (1984) *Improving Leadership Effectiveness: The Leader Match Concept*, 2nd ed. New York: Wiley.

FISKE, SUSAN T., AND SHELLEY E. TAYLOR (1984) *Social Cognition*. Reading, MA: Addison-Wesley.

FREEDMAN, LAWRENCE (1997) "How Did the Democratic Process Affect Britain's Decision to Reoccupy the Falkland Islands?" In *Paths to Peace: Is Democracy the Answer?* edited by Miriam Fendius Elman. Cambridge, MA: MIT Press.

FREEDMAN, ROBERT O. (1982) *Soviet Policy Toward the Middle East Since 1970*. New York: Praeger.

FREIDMAN, WILLIAM (1994) Woodrow Wilson and Colonel House and Political Psychobiography. *Political Psychology* **15**:35–59.

GARBA, JOE (1987) *Diplomatic Soldiering: Nigerian Foreign Policy, 1975–1979*. Ibadan, Nigeria: Spectrum.

GARDNER, HOWARD (1983) *Frames of the Mind: The Theory of Multiple Intelligences*. New York: Basic Books.

GEORGE, ALEXANDER L. (1980) *Presidential Decision Making in Foreign Policy: The Effective Use of Information and Advice*. Boulder, CO: Westview Press.

GEORGE, ALEXANDER L., AND ERIC K. STERN (1998) "Presidential Management Styles and Models." In *Presidential Personality and Performance*, edited by Alexander L. George and Juliette L. George. Boulder, CO: Westview Press.

GEYER, D. (1988) *Castro: A Political Biography*. New York: New York Times Books.

GLAD, BETTY (1996) Passing the Baton: Transformational Political Leadership from Gorbachev to Yeltsin, from de Klerk to Mandela. *Political Psychology* **17**:1–28.

GOFFMAN, E. (1959) *The Presentation of Self in Everyday Life*. Garden City, NY: Doubleday.

GOLDMAN, ERIC F. (1969) *The Tragedy of Lyndon Johnson*. New York: Alfred A. Knopf.

GREENSTEIN, FRED I. (1987) *Personality and Politics*. Princeton, NJ: Princeton University Press.

GUNTHER, RICHARD (1988) *Politics and Culture in Spain*. Politics and Culture Series no. 5. Ann Arbor: Institute for Social Research, University of Michigan.

HAGAN, JOE D. (1994) Domestic Political Systems and War Proneness. *Mershon International Studies Review* **38**:183–208.

HAGAN, JOE D. (1995) "Domestic Political Explanations in the Analysis of Foreign Policy." In *Foreign Policy Analysis*, edited by Laura Neack, Jeanne A. K. Hey, and Patrick J. Haney. Englewood Cliffs, NJ: Prentice-Hall.

HALL, RICHARD L., AND ROBERT P. VAN HOUWELING (1995) Avarice and Ambition in Congress: Representatives' Decisions to Run or Retire from the U. S. House. *American Political Science Review* **89**:121–136.

HAMPSON, FEN O. (1988) "The Divided Decision-Maker: American Domestic Politics and the Cuban Crisis." In *The Domestic Sources of American Foreign Policy*, edited by Charles W. Kegley, Jr., and Eugene R. Wittkopf. New York: St. Martin's Press.

HARGROVE, ERWIN C. (1989) "Two Conceptions of Institutional Leadership." In *Leadership and Politics: New Perspectives in Political Science*, edited by Bryan D. Jones. Lawrence: University of Kansas Press.

'T HART, PAUL (1990) *Groupthink in Government: A Study of Small Groups and Policy Fiascoes*. Amsterdam: Swetz and Zeitlinger.

HEIKAL, MOHAMED H. (1975) *The Road to Ramadan*. Glasgow: William Collins.

HERMANN, CHARLES F. (1972) *International Crises: Insights from Behavioral Research*. New York: Free Press.

HERMANN, MARGARET G. (1980) Explaining Foreign Policy Behavior Using the Personal Characteristics of Political Leaders. *International Studies Quarterly* **24**:7–46.

HERMANN, MARGARET G. (1984) "Personality and Foreign Policymaking: A Study of 53 Heads of Government." In *Foreign Policy Decision Making: Perceptions, Cognitions, and Artificial Intelligence*, edited by Donald A. Sylvan and Steve Chan. New York: Praeger.

HERMANN, MARGARET G. (1988a) "The Role of Leaders and Leadership in the Making of American Foreign Policy." In *The Domestic Sources of American Foreign Policy*, edited by Charles W. Kegley, Jr., and Eugene R. Wittkopf. New York: St. Martin's Press.

HERMANN, MARGARET G. (1988b) "Hafez al-Assad, President of Syria: A Leadership Profile." In *Leadership and Negotiation: A New Look at the Middle East*, edited by Barbara Kellerman and Jeffrey Rubin. New York: Praeger.

HERMANN, MARGARET G. (1993) "Leaders and Foreign Policy Decision Making." In *Diplomacy, Force, and Leadership*, edited by Dan Caldwell and Timothy J. McKeown. Boulder, CO: Westview Press.

HERMANN, MARGARET G. (1995) Leaders, Leadership, and Flexibility: Influences on Heads of Government as Negotiators and Mediators. *The Annals of the American Academy of Political and Social Science* **542**:148–167.

HERMANN, MARGARET G. (1999a) *Assessing Leadership Style: A Trait Analysis*. Columbus, OH: Social Science Automation, Inc. (Available at www.socialscienceautomation.com)

HERMANN, MARGARET G. (1999b) *Leadership Profile of Mohammad Khatami.* Columbus, OH: Social Science Automation, Inc.

HERMANN, MARGARET G. (1999c) *Leadership Profile of Slobodan Milosevic.* Columbus, OH: Social Science Automation, Inc.

HERMANN, MARGARET G. (1999d) *Leadership Profile of Zoran Djindjic.* Columbus, OH: Social Science Automation, Inc.

HERMANN, MARGARET G., AND JOE D. HAGAN (1998) International Decision Making: Leadership Matters. *Foreign Policy* **100** (spring):124–137.

HERMANN, MARGARET G., AND CHARLES F. HERMANN (1982) "A Look Inside the Black Box: The Need for a New Area of Inquiry." In *Biopolitics, Political Psychology, and International Politics*, edited by Gerald W. Hopple. New York: St. Martin's Press.

HERMANN, MARGARET G., AND CHARLES F. HERMANN (1989) Who Makes Foreign Policy Decisions and How: An Empirical Inquiry. *International Studies Quarterly* **33**:361–387.

HERMANN, MARGARET G., AND CHARLES W. KEGLEY, JR. (1995) Rethinking Democracy and International Peace: Perspectives from Political Psychology. *International Studies Quarterly* **39**:511–533.

HERMANN, MARGARET G., AND THOMAS PRESTON (1994) Presidents, Advisers, and Foreign Policy: The Effect of Leadership Style on Executive Arrangements. *Political Psychology* **15**:75–96.

HERMANN, MARGARET G., AND THOMAS PRESTON (1999) "Presidents, Leadership Style, and the Advisory Process." In *Domestic Sources of American Foreign Policy*, edited by James McCormick and Eugene R. Wittkopf. Lanham, MD: Rowman and Littlefield.

HOOPES, TOWNSEND (1969) *The Limits of Intervention.* New York: David McKay.

INEGBEDION, E. JOHN (1991) "Nigerian Policy in Southern Africa: Frontline State or Rearguard Actor." In *Prospects for Peace and Development in Southern Africa in the 1990s: Canadian Comparative Perspectives*, edited by Larry W. Swatuk and Timothy M. Shaw. Lanham, MD: University Press of America.

JANIS, IRVING L. (1972) *Victims of Groupthink: A Psychological Study of Foreign Policy Decisions and Fiascoes.* Boston: Houghton Mifflin.

JOHNSON, LYNDON BAINES (1971) *The Vantage Point: Perspectives of the Presidency, 1963–1969.* New York: Holt, Rinehart, and Winston.

JOHNSON, RICHARD T. (1974) *Managing the White House: An Intimate Study of the Presidency.* New York: Harper and Row.

JONSSON, CHRISTER (1982) *Cognitive Dynamics and International Politics.* New York: St. Martin's Press.

KAARBO, JULIET, AND MARGARET G. HERMANN (1998) Leadership Styles of Prime Ministers: How Individual Differences Affect the Foreign Policymaking Process. *Leadership Quarterly* **9**:243–263.

KEARNS, DORIS (1976) *Lyndon Johnson and the American Dream.* New York: Harper and Row.

KHONG, YUEN FONG (1992) *Analogies at War: Korea, Munich, Dien Bien Phu, and the Vietnam Decisions of 1965.* Princeton, NJ: Princeton University Press.

KIRK-GREENE, ANTHONY, AND DOUGLAS RIMMER (1981) *Nigeria Since 1970: A Political and Economic Outline.* London: Hodder and Stoughton.

KISSINGER, HENRY A. (1994) *Diplomacy.* New York: Simon and Schuster.

KORANY, BAGHAT (1986a) When and How Do Personality Factors Influence Foreign Policy? A Comparative Analysis of Egypt and India. *Journal of South Asian and Middle Eastern Studies* **9**:35–59.

KORANY, BAGHAT (1986b) *How Foreign Policy Decisions Are Made in the Third World.* Boulder, CO: Westview Press.

KORANY, BAGHAT (1990) Inferior Capabilities and War Decisions: The Interaction of the Predominant Leader and Environmental Pressures in Egypt, October 1973. Paper presented at the annual meeting of the International Studies Association, Washington, D.C., April.

KOTTER, JOHN P., AND PAUL R. LAWRENCE (1974) *Mayors in Action.* New York: Wiley.

KOWERT, PAUL, AND MARGARET G. HERMANN (1997) Who Takes Risks? Daring and Caution in Foreign Policy Making. *Journal of Conflict Resolution* **41**:611–637.

KUPCHAN, CHARLES A. (1994) *The Vulnerability of Empire.* Ithaca, NY: Cornell University Press.

LAU, RICHARD R., AND DAVID O. SEARS (1986) *Political Cognition.* Hillsdale, NJ: Lawrence Erlbaum.

LEBOW, RICHARD NED (1981) *Between Peace and War: The Nature of International Crisis.* Baltimore, MD: Johns Hopkins University Press.

LEVY, JACK S. (1989) "The Causes of War: A Review of Theories and Evidence." In *Behavior, Society, and Nuclear War*, vol. 1, edited by Philip E. Tetlock, Jo L. Husbands, Robert Jervis, P. S. Stern, and Charles Tilly. New York: Oxford University Press.

LEWIN, KURT, AND RONALD LIPPIT (1938) An Experimental Approach to the Study of Autocracy and Democracy: A Preliminary Note. *Sociometry* **1**:292–300.

LINDSAY, P. H., AND D. NORMAN (1977) *Human Information Processing*. New York: Academic Press.

MANLEY, JOHN F. (1969) Wilbur D. Mills: A Study in Congressional Influence. *American Political Science Review* **63**:442–464.

MCCLELLAND, DAVID D. (1975) *Power: The Inner Experience*. New York: Irvington Publishers.

MCGRATH, JOSEPH E. (1984) *Groups: Interaction and Performance*. Englewood Cliffs, NJ: Prentice-Hall.

MCNAMARA, ROBERT S. (1995) *In Retrospect: The Tragedy and Lessons of Vietnam*. New York: Random House.

MCPHERSON, HARRY C. (1972) *A Political Education*. Boston: Little, Brown.

MCPHERSON, HARRY C. (1995) Interview with Thomas Preston, July 7.

MERRITT, RICHARD L., AND DINA A. ZINNES (1991) "Democracies and War." In *Measuring War*, edited by Alex Inkeles. New Brunswick, NJ: Transaction Books.

NEUMANN, ROBERT G. (1983–84) Assad and the Future of the Middle East. *Foreign Affairs* **62**:240–256.

*New York Times* (1990) Cheney Steps to Center of the Lineup. August 24, 3.

NIXON, RICHARD M. (1982) *Leaders*. New York: Warner Books.

OGUNSANWO, ALABA (1980) *The Nigerian Military and Foreign Policy, 1975–1979: Process, Principles, Performance, and Contradictions*. Princeton, NJ: Center for International Relations, Princeton University.

PAKRADOUNI, KARIM (1983) Hafez al-Assad—The Arabs' Bismarck. *Manchester Guardian Weekly*, vol. 129, December 11, 14.

PAYNE, J., OLIVER WOSHINSKY, E. VEBLEN, W. COOGAN, AND G. BIGLER (1984) *The Motivation of Politicians*. Chicago: Nelson-Hall.

PETTY, RICHARD E., AND JOHN T. CACIOPPO (1986) "The Elaboration Likelihood Model of Persuasion." In *Advances in Social Psychology*, vol. 19, edited by Leonard Berkowitz. New York: Academic Press.

POWELL, COLIN (1995) *My American Journey*. New York: Random House.

PRESTON, THOMAS (1996) The President and His Inner Circle: Leadership Style and the Advisory Process in Foreign Policy Making. Ph.D. dissertation, Ohio State University.

PRESTON, THOMAS (2000) "The President's Inner Circle: Personality and Leadership Style in Foreign Policy Decisionmaking." In *Presidential Power: Forging the Presidency for the Twenty-First Century*, edited by Robert Y. Shapiro, Martha Joynt Kumar, and Lawrence R. Jacobs. New York: Columbia University Press.

PRESTON, THOMAS (2001) *The President and His Inner Circle: Leadership Style and the Advisory Process in Foreign Policy Making.* New York: Columbia University Press.

PUTNAM, ROBERT D. (1988) Diplomacy and Domestic Politics: The Logic of Two-Level Games. *International Organization* **42**:427–460.

QUANDT, WILLIAM B. (1977) *Decade of Decisions: American Policy Toward the Arab-Israeli Conflict 1967–1976.* Berkeley: University of California Press.

RIPLEY, BRIAN, AND JULIET KAARBO (1992) Lyndon Johnson and the 1965 Decision to Escalate U.S. Involvement in Vietnam. Manuscript, Mershon Center, Ohio State University.

ROCKMAN, BERT A. (1991) "The Leadership Style of George Bush." In *The Bush Presidency First Appraisals*, edited by Colin Campbell and Bert A. Rockman. Chatham, NJ: Chatham House.

RUBINSTEIN, ALVIN Z. (1977) *Red Star on the Nile.* Princeton, NJ: Princeton University Press.

RUSK, DEAN (1969) Oral History Interview, July 28.

SAFRAN, NADAV (1989) "Dimensions of the Middle East Problem." In *Foreign Policy in World Politics: States and Regions*, 7th ed., edited by Roy C. Macridis. Englewood Cliffs, NJ: Prentice-Hall.

SCHRODT, PHILIP A., AND DEBORAH J. GERNER (2000) Cluster-Based Early Warning Indicators for Political Change in the Contemporary Levant. *American Political Science Review* **94**:803–817.

SCHRODT, PHILIP A., SHANNON G. DAVIS, AND JUDITH L. WEDDLE (1994) Political Science: KEDS—A Program for the Machine Coding of Event Data. *Social Science Computer Review* **12**:561–588.

SHAW, TIMOTHY M. (1987) Nigeria Restrained: Foreign Policy Under Changing Political and Petroleum Regimes. *The Annals of the American Academy of Political and Social Science* **489**:40–50.

SHAW, TIMOTHY M., AND OLAJIDE ALUKO (1983) *Nigerian Foreign Policy: Alternative Perceptions and Projections.* London: Macmillan.

SHAZLY, SAAD AL (1980) *The Crossing of the Suez.* San Francisco: American Mideast Research.

SHEPARD, ROBERT B. (1991) *Nigeria, Africa, and the United States from Kennedy to Reagan.* Bloomington: Indiana University Press.

SNARE, CHARLES E. (1992) "Applying Personality Theory to Foreign Policy Behavior: Evaluating Three Methods of Assessment." In *Political Psychology and Foreign Policy*, edited by Eric Singer and Valerie M. Hudson. Boulder, CO: Westview Press.

SNYDER, JACK (1991) *Myths of Empire: Domestic Politics and International Ambition*. Ithaca, NY: Cornell University Press.

SNYDER, MARK (1987) *Public Appearances, Private Realities: The Psychology of Self-Monitoring*. New York: W. H. Freeman.

STEIN, JANICE GROSS (1994) Political Learning by Doing: Gorbachev as Uncommitted Thinker and Motivated Learner. *International Organization* **48**:155–183.

STEINBRUNER, JOHN D. (1974) *The Cybernetic Theory of Decision: New Dimensions of Political Analysis*. Princeton, NJ: Princeton University Press.

STEWART, PHILIP D., MARGARET G. HERMANN, AND CHARLES F. HERMANN (1989) Modeling the 1973 Soviet Decision to Support Egypt. *American Political Science Review* **83**:35–59.

STOESSINGER, JOHN D. (1979) *Crusaders and Pragmatists: Movers of Modern American Foreign Policy*. New York: W. W. Norton.

SUEDFELD, PETER (1992) Cognitive Misers and Their Critics. *Political Psychology* **13**:435–453.

SUEDFELD, PETER, AND MICHAEL D. WALLACE (1995) "President Clinton as a Cognitive Manager." In *The Clinton Presidency: Campaigning, Governing, and the Psychology of Leadership*, edited by Stanley A. Renshon. Boulder, CO: Westview Press.

SUEDFELD, PETER, PHILIP E. TETLOCK, AND S. STREUFERT (1992) "Conceptual/Integrative Complexity." In *Motivation and Personality: Handbook of Thematic Content Analysis*, edited by C. P. Smith. Cambridge: Cambridge University Press.

TETLOCK, PHILIP E. (1991) An Integratively Complex Look at Integrative Complexity. Paper presented at the annual meeting of the American Psychological Association, San Francisco, August.

THOMSON, JAMES C., JR. (1968) How Could Vietnam Happen? An Autopsy. *The Atlantic Monthly*, April.

VANCE, CYRUS (1983) *Hard Choices: Critical Years in America's Foreign Policy*. New York: Simon and Schuster.

VROOM, V. H., AND P. H. YETTON (1973) *Leadership and Decision Making*. Pittsburgh, PA: University of Pittsburgh Press.

WALKER, STEPHEN G. (1983) The Motivational Foundations of Political Belief Systems: A Re-Analysis of the Operational Code Construct. *International Studies Quarterly* **27**:179–201.

WINTER, DAVID G. (1992) "Personality and Foreign Policy: Historical Overview." In *Political Psychology and Foreign Policy*, edited by Eric Singer and Valerie M. Hudson. Boulder, CO: Westview Press.

WINTER, DAVID G. (1995) "Presidential Psychology and Governing Styles: A Comparative Psychological Analysis of the 1992 Presidential Candidates." In *The Clinton Presidency: Campaigning, Governing, and the Psychology of Leadership*, edited by Stanley A. Renshon. Boulder, CO: Westview Press.

WINTER, DAVID G., MARGARET G. HERMANN, WALTER WEINTRAUB, AND STEPHEN G. WALKER (1991) The Personalities of Bush and Gorbachev Measured at a Distance: Procedures, Portraits, and Policy. *Political Psychology* **12**:215–243.

WOODWARD, BOB (1991) *The Commanders*. New York: Simon and Schuster.

WOSHINSKY, OLIVER H. (1973) *The French Deputy: Incentives and Behavior in the National Assembly*. Lexington, MA: D. C. Heath.

YOUNG, MICHAEL D. (2000) Automating Assessment at a Distance. *The Political Psychologist* **5**:17–23.

YOUNG, MICHAEL D., AND MARK SCHAFER (1998) Is There Method in Our Madness? Ways of Assessing Cognition in International Relations. *Mershon International Studies Review* **42**:63–96.

ZILLER, ROBERT C. (1973) *The Social Self*. New York: Pergamon.

ZILLER, ROBERT C., WILLIAM F. STONE, ROBERT M. JACKSON, AND NATALIE J. TERBOVIC (1977) "Self-Other Orientations and Political Behavior." In *A Psychological Examination of Political Leaders*, edited by Margaret G. Hermann. New York: Free Press.

132

# Resolve, Accept, or Avoid:

## Effects of Group Conflict on Foreign Policy Decisions

*Charles F. Hermann*

Bush School, Texas A&M University

*Janice Gross Stein*

University of Toronto

*Bengt Sundelius*

Uppsala University

*Stephen G. Walker*

Arizona State University

Groups are pervasive decision units in governments. Legislative committees, cabinets, military juntas, politburos of ruling parties, and executive councils are all candidates. The operation of many government ministries and agencies suggests that groups are also frequently at the core of the bureaucratic process. Coordination between bureaucracies creates both ad hoc and standing interdepartmental committees and boards that serve as decision units. In governments, groups usually convene to cope with problems. Such policy problems typically involve complex, cognitive tasks with no single correct answer.

If no one individual alone has the authority to act on behalf of a government, then we must turn to alternative decision units. As we have just observed, another possible, and frequently encountered, configuration is the single group. By a single group we mean an entity of two or more people all of whom interact

© 2001 International Studies Association
Published by Blackwell Publishers, 350 Main Street, Malden, MA 02148, USA, and 108 Cowley Road, Oxford OX4 1JF, UK.

directly with one another and collectively reach a decision. No definite boundaries are proposed with respect to the upper limit of group size. Thus, the group may be as small as two or three people or as large as a parliament of hundreds, so long as there is a collective, interactive decision process in which all the members who are needed to make authoritative commitments participate. (In practice, however, many large groups subdivide into committees, coalitions, or other subsets to conduct much of their decision making.) There is no stipulation in our definition that the group be a formal or legal body. For a group to be authoritative, it must have the definitive ability to commit or withhold the relevant governmental resources on the subject matter of the decision even if the entity is ad hoc or not part of an established institutional structure. (The ability to commit or withhold resources does not mean that group members themselves will actually implement the decision, leaving open potential discrepancies between choice and action.)

In differentiating a single group from the predominant leader and coalition decision units, we recognize several boundary issues. When a strong leader operates with a group of advisers, we may have difficulty determining whether the unit is a predominant leader or a single group. So long as the leader alone has the power to commit the regime's resources and does not delegate formally or tacitly that decision to advisers, the unit is a predominant leader. Another potential ambiguity arises when, for example, a parliamentary government consists of a multiparty coalition cabinet. In this case, the distinction must be made between the single group and coalition of autonomous actors whose representatives may meet together. When individual cabinet members are bound to specific positions taken elsewhere (e.g., by their political party) and they are not free to act independently, then the authoritative decision unit is a coalition. If, however, the cabinet officers can form or change their positions on a problem without outside consultation, then the unit is a single group.[1]

---

[1] The conceptual boundaries between single group and either predominant leader or coalition decision units used in this special issue establish useful theoretical differentiations that facilitate inquiry. The boundaries allow us to isolate group effects on decision making. Nonetheless, it may be possible to conceive of a group effect under circumstances defined here in the domain of a predominant leader or coalition. For example, Garrison (1999) and Preston (2001) have explored the influence of U.S. presidents and their advisers on foreign policy, that is, leader-group interactions. A broader definition of the single group decision unit might state that if a group is present throughout the deliberations with a predominant leader, then the decision may reflect a collective effort. Similarly, if representatives of multiple organizations (i.e., a coalition) continuously meet to reach a decision, then the result may be more readily described as made by a group. For further discussion of this point see the last article in this special issue which presents an analysis of sixty-five cases using the decision units framework.

Through this examination of single groups, we seek to explore the following questions: How do group dynamics affect the choice or nature of a decision in foreign policy? When disagreements within a group arise, how are these conflicts handled? Clearly, there are many sources of influence on any foreign policy decision. The immediate concern in this essay is with the potential impact of structure and process on the output when the decision unit is a group.

## Managing Group Conflict

### *Information and Options*

The study of decision making in problem-solving groups suggests two major categories of activities essential to choice—the processing of information and the management of options. In the actual performance of groups information and option management are intertwined. For analytical purposes, however, it is useful to distinguish between them. Information management concerns the array of issues associated with the nature, structuring, and dissemination of information within a group. Option management concerns the development, advocacy, assessment, and selection of an option or alternative. An option or alternative is an expressed means of treating or coping with a recognized problem including as one possibility doing nothing. A great deal of group research has been conducted on communications and information management. Semmel (1982), Vertzberger (1990), and Gaenslen (1992) have each provided interpretive summaries of research on information processing as it applies to political groups.

Rather than focus on information processing, the models developed in this essay build upon option management in groups and, particularly, on the role that conflict among group members plays in that process. If from the beginning of their deliberations all members of a group prefer the same method for dealing with a problem and a specific means of implementing that solution, group decision making is relatively simple and straightforward. In dealing with policy problems in a political setting, however, members frequently disagree on preferred modes of coping with a problem. As has often been noted, conflict and its management are the essence of politics and policymaking.

### *Group Decision Outputs*

A group's management of substantive disagreements among its members affects the resulting output, that is, their collective "solution" to the problem. That assumption is the cornerstone of this inquiry. And to begin with, we must construct a broadly applicable and meaningful way to categorize a group's decision outputs.

How might the options or choices of group problem solving be conceptualized? In some cases, a group may face a dichotomous choice—a jury must decide guilt or innocence; a commission must vote yes or no on a resolution, a cabinet has to choose whether or not to continue present policy. Other problems offer choice along a single dimension, usually involving a numerical continuum. For instance, how many dollars should be allocated to a department's budget; what percentage of the delegates should be selected by a primary; how many of the potential targets for aerial bombing should be selected? Still other problems may be "packaged" such that the choice is among multiple dimensions at the same time—should the organization be configured with three or four departments with some units headed by directors and others by ministers or should all heads of the departments be of the same rank, for example? Finally, the choices for dealing with some problems may appear to be framed without respect to any underlying common structure or dimensions. Consider that in the 1962 Cuban missile crisis, American policymakers were deciding among air strikes, an invasion, or a naval blockade.

The ways in which options are defined influences the effects that group processes can have on the output. For example, if the disagreements among group members are conceptualized as choice points on continuous dimensions, then analysis might involve indifference curves and trade-offs. Alternatively, studies exploring whether groups using a majority decision rule make different decisions from those requiring unanimity show that the results vary depending on whether the choice is dichotomous (as in a jury trial) or among multiple options along a continuum (as in budget allocations) (for reviews of this literature see Miller, 1985, and Hastie, 1993).

In the present essay, the focus is on group choices among one or more options lacking any perceived common underlying structure. Unstructured solutions are specified as the outcome to be explained for several reasons. First, a large number of public policy problems appear to have this quality. Indeed, outside of budget questions, exactly such problems are among the most complex and demanding choices policymakers face. Second, we are interested in how decision makers resolve internal disagreements. Thus, we want to minimize the explanations that rest on the possibility of "prominent solutions" such as splitting the difference or that lead to a mathematical best solution, for example, a pareto optimal choice. We are interested in policy problems that lack a determinable "right answer."

More specifically, the unstructured solutions that constitute the dependent variables in this work can be differentiated as: (a) deadlock; (b) prevalent solution; (c) subset solution; and (d) integrative solution. *Deadlock* defines a situation of stalemate in which group members reach no decision on how to resolve their differences. *Dominant* solution refers to a situation in which the group selects the one option that has been discussed from the outset; they exhibit concurrence around a particular choice. In this type of situation it is a choice between that option or doing nothing. Such circumstances may result when no other option is perceived to meet the apparent decision criteria, when norms prevent articulating an alternative to an option advocated by an authoritative group member, or when there is a shared set of beliefs about what is appropriate in the particular context.[2] A *subset* solution is one that is satisfactory to some faction in the group, but does not capture the preferences of all members. Subset solutions generally mean that a particular faction's position prevails and takes priority over other members and factions' preferences in the group's decision. Finally, an *integrative* solution is an alternative that in the course of the group discussion comes to represent the preference of all members and involves some shift from their initial choices. An integrative solution may result from successful persuasion of some members by others to change their explicitly stated preference, by a shift in the preference orderings of all members, perhaps as a result of the creation of a new option not initially recognized by the group, or through achieving a mutually acceptable compromise.

Although these unstructured solutions are nominal categories, under some circumstances they may be viewed as representing an ordinal scale. To create such a scale requires us to assume that deadlock is the preference of no group member and that the dominant solution is the real preference of only one or two group members—perhaps an authoritarian leader or powerful faction. If that is assumed, then the outcomes are a continuum representing the number of group members whose preferences are captured in the solution. Deadlock is no one's preference. Dominant solutions represent one (or two) members' preferences. A subset solution is a coalition of members' preferences (more than one or two, but less than all). And an integrative solution captures the final preference of all members. Beyond this possible ordering of the solutions, they can reflect something about the quality of group decision making. We will return to this point later.

## *Managing Conflict*

Opposing difficulties can beset a group's management of options. On the one hand, groups can lock in on a single option (sometimes flowing from an uncontested definition of the problem) and accept it, perhaps, uncritically. On the

---

[2] Mintz and his associates (1997) have advanced a "poliheuristic theory of decision" that recognizes policymakers may categorically reject options that fail to meet some strongly held value for which they accept no compensatory trade-off, even if such options have other highly desirable properties. If all but one option fail to fulfill such a requirement, then decision makers may conclude there is only one possible approach to a problem.

other hand, groups may be divided among several competing options and be unable to resolve their differences and achieve closure, or they may follow a procedure for resolving the differences that makes little reference to the relative merits of the advocated options.

We propose three basic models regarding how decision-making groups consider options drawn from insights about how group members manage substantive disagreement among themselves over preferred options. Essentially, the models represent three different ways groups can deal with internal disagreement or conflict. They can avoid it, they can resolve it, or they can accept it. The three models are designated:

> Concurrence (producing a tendency to *avoid* group conflict)
>
> Unanimity (producing a tendency to *resolve* group conflict)
>
> Plurality (producing a tendency to *accept* group conflict)

The basic argument of this essay is that each of these modes of dealing with substantive disagreement in problem-solving groups creates a tendency toward one of the four solutions described above, namely, deadlock, dominant, subset, or integrative. To demonstrate their operation and plausibility, the models will be illustrated using three case studies: (1) the British cabinet decisions in the 1938 Munich crisis; (2) the Swedish government's responses to the discovery in 1981 that a Soviet submarine had run aground on their coast; and (3) the Israeli decisions in reaction to the 1967 Egyptian offensive preparations for an apparent military attack. But before we turn to explicating the models and the cases, let us examine previous theoretical work done on decision making in political groups which provides the foundation for proposing the hypothesized effects of a group's conflict strategy on its decisions.

## *Theoretical Foundations*

Three major theoretical perspectives provide the conceptual grounding for the proposed models. In each case particular features have been selected or elaborated in the current presentation.[3] Groupthink, advanced by Janis (1972, 1982), provides the touchstone for the concurrence model. With its emphasis on suppression of disagreement to preserve the well-being of the decision unit, groupthink offers insights into conditions that can produce what we describe as dominant solutions.

---

[3] It is important to underscore that in the discussion of each model in this essay selected features of the theoretical frameworks are emphasized without any attempt to provide a comprehensive interpretation of the original work.

Bureaucratic politics associated with the scholarship of a number of individuals including Huntington (1960), Allison (1971), and Halperin (1974) can be used to establish the foundation for the unanimity model and, to some degree, the plurality model as well.[4] The struggles among group members advocating the preferences of their respective agencies, which lies at the core of bureaucratic politics, provide us with quite different group problem-solving processes than those suggested by the groupthink perspective. The conflict envisioned by bureaucratic politics produces varying dynamics depending upon the group's established practice for handling recognized disagreements. If there is no formal decision rule or group norm for resolving such conflicts, then the solution is likely deadlock. Explicit decision rules or norms can provide a means for creating other probable decision outputs.

The use of a group norm or formal decision rule as an accepted procedure for resolving disagreements within a group taps into another body of research. Whether in a jury or some other decision group, the use of a plurality or majority rule accepts the possibility that a collective decision can be reached even though some members of the group continue to disagree with that choice. We view this process as leading to the acceptance of conflict as described by the plurality model. By contrast, when a group operates with a unanimity rule or norm, then the members must somehow resolve their disagreements to the satisfaction of all members if any decision is to occur. Research in social psychology suggests different processes and probable results depending on which decision rule prevails (e.g., Thompson, Mannix, and Bazerman, 1988; Miller, 1989; Kameda and Sugimori, 1993; Nielson and Miller, 1997). Thus, the study of groups using different decision rules can help us refine the expectations of bureaucratic politics, creating the basis for distinguishing between the Unanimity Model that approximates a strict form of bureaucratic politics and the Plurality Model in which continued disagreement by at least some members is acceptable. The reasons for these expectations are set forth below.

## CONCURRENCE MODEL

### Avoiding Conflict

When problems are complex, uncertainty abounds, and the stakes are high—as is often the case with policy problems, it is not surprising that individuals frequently differ on what should be done. Disagreement seems more likely when the participants have significantly different life experiences and political or

---

[4] For an extensive contemporary review of the bureaucratic politics literature see the pieces in the symposium edited by Stern and Verbeek (1998).

administrative responsibilities. Although substantive disagreement in groups may be common, it can be debilitating. Humans frequently have difficulty distinguishing between another person's critique of their position and an attack on them as a person. (In fact, sometimes in policy circles the purpose of policy disagreement is to discredit the person holding opposing views.) Members may devote their energies to acrimonious attacks and rebuttals, to distortions or the withholding of information, to the creation of opposing subgroups, and a variety of other activities that not only prevent effective problem management but also can destroy the group as a collective entity.

If members value their group, for whatever reason, and if they are concerned about the potential damaging consequences of conflict, then they may act so as to minimize any substantive disagreements among themselves. Individuals and cultures may also attach high value to smooth, tranquil interpersonal relations. To preserve harmony in the valued group, members may suppress disagreements and engage in efforts to promote concurrence on substantive issues. This basic insight lies at the core of groupthink as developed by Janis (1972, 1982).

Many interpretations and evaluations of the groupthink concept have been advanced.[5] The Concurrence Model being described here makes very selective use of the ideas associated with groupthink. Our initial task is to sketch the likely processes that can occur in a group determined to avoid internal disagreements and to hypothesize the probable decision outcomes.

The process usually associated with how a group will manage a policy problem when under the influence of groupthink is premature closure around an initially advocated course of action. A group experiences premature closure when it accepts the option prominently presented, usually by an authoritative member, early in its deliberations without engaging in a serious evaluation of its potential limitations or undertaking a careful comparison of it with any other possible alternatives. In advancing the framework for groupthink, Janis (1982) considered a number of antecedent conditions that he argued contributed to the process of prompt concurrence-seeking and premature closure. Among these were situational properties (e.g., high stress) and structural features (e.g., common social and ideological backgrounds among members), but he contended that the one necessary condition was substantial group cohesion. "Only when a group of policymakers is moderately or highly cohesive can we expect the groupthink syndrome to emerge" (Janis, 1982:176). Earlier he noted (Janis, 1972:13): "*The more amicability and esprit de corps among members of a policymaking in-group, the greater is the danger that independent critical think-*

---

[5] See, e.g., Longley and Pruitt (1980), McCauley (1989), and 't Hart, Stern, and Sundelius (1997).

*ing will be replaced by groupthink"* (emphasis in original). In concluding his original treatise on groupthink, Janis (1972:197) observed: "The prime condition repeatedly encountered in case studies of fiascoes is group cohesiveness."[6]

## *Primary Group Identity*

Group cohesion has long been recognized as a theoretically powerful variable in sociology and social psychology. Summarizing the research of others on cohesion, McGrath (1984:241) reports it is "the sum total of all the forces attracting members to a group." These forces are usually seen as the positive attraction among group members for each other, the value members attach to the work the group does, and the prestige or status individuals perceive to be associated with belonging to the group.

Although group cohesion provides an important explanatory variable in considerations of groupthink, it does not capture the idea of *relative* attraction that occurs when individuals are simultaneously in several groups or organizations dealing with the same policy problem. For example, in government, an individual may be involved in an interagency task force as a representative of some bureau or organization. The person may be attracted to the interagency task force (because of its prestige or the talented people it includes), but remain committed to his or her home department. Members of the task force, even though they greatly admire others in the group (i.e., manifest high affiliation needs and display cohesion), may feel an even stronger attachment to their home department since it determines their promotions, merit pay raises, and other rewards.

We need a concept that captures the member's attraction to a particular group relative to his or her commitment to other associations working on the policy problem. We propose the concept of primary group identity to capture this idea. In other words, when faced with conflicting pressures from entities with which the individual member identifies, does that group member attach a higher positive value to his or her present decision group membership or to some other outside association? A member's attraction to the group needs to be not only positive (group cohesion) but also more valued than other relevant identities with respect to preferences on the immediate policy problem.

It is this notion of whether, in situations of competing demands, members' overriding commitment is to the authoritative decision group that provides a

---

[6] George (1997) interprets Janis as proposing that concurrence-seeking in groups occurs only as a result of severe externally produced stress that, in turn, triggers an increased need among members for affiliation (i.e., a strong need for cohesion). The interpretation advanced here is that a strong need to avoid disruptive conflict in the group can have the same consequence.

critical key for distinguishing group processes. When the primary identity of most group members is to the group, by definition they will place a high value on its continuation and well-being. They also will want to preserve their own good standing in the group. This commitment creates a predisposition in members to guard against activity that (1) may threaten the group, such as divisive disputes among members, or (2) may reduce their own acceptance by the other members—such as by expressing views that appear to be at odds with those of others in the group. In short, to preserve the well-being of the group, members will tend to avoid disagreements over the substantive problems they are addressing. By avoiding conflict, however, they risk reducing the critical assessment of ideas presented to them.

If strong group identity triggers the inhibiting processes just discussed (that is, avoiding disagreement and uncritically supporting the emerging course of action), then premature closure of the kind envisioned in groupthink is highly likely. It is probable that the group will adopt the dominant solution. This solution will be proposed early in the group's deliberations and promptly endorsed by key members after which no other options are likely to be seriously considered. The resulting action will be less complex, unqualified, and, in some sense, more extreme than an option chosen after rigorous evaluation and comparison to other alternatives. Of course, the dominant course of action presented to group members may turn out to be balanced and sensitive to nuances in the current situation. But without discussion of competing alternatives or criticism of the initial proposal, contextual subtleties and possibilities not anticipated by the original proposer are less likely to be taken into account. It is precisely this type of group interaction that is suppressed in the Concurrence Model for fear of creating interpersonal conflict and damage to the group's well-being.[7]

The reader should recognize that the effect is the same even if members of the group actually agree that the initially proposed solution is the only feasible option under the present circumstances and engage in no critical assessment because they have privately concluded it is the best (or only) course of action. Moreover, the outcome is the same as when some members privately question the merits of the proposed solution, but in the interest of group harmony do not speak out. Aware of this danger, George (1972) advocated building into the process what he called multiple advocacy. In his prescription, advocates of

---

[7] If groups permit disagreement and debate among members over the merits of a proposed option, the reverse effect can occur; that is, an initially advocated, coherent option can be diminished in quality by the modifications added to accommodate the preferences of other members. This danger may be most likely when agreement of all members is required for action.

different approaches are deliberately built into the discussion to avoid premature acceptance of an option. Such a procedure, he believed, would help to avoid quick group acceptance for the reasons stated above.

## *Concurrence Mediating Variables*

A moment's reflection might lead one to the conclusion that a group whose members give it their primary identity would be exactly the collectivity where individuals would feel free to disagree. If they are comfortable with one another and have a good working relationship, dissent may not generate the personal antagonism or fear of group upheaval that disagreement might generate under other conditions. If such is the case, then we would expect other additional considerations to be important in determining whether primary group identity leads to suppression or acceptance of substantive disagreement. Specifically, group norms about the permissibility of expressing disagreement may be critical. Even if the group lacks such shared ways of behaving but has a current leader who allows disagreement, then premature closure can be avoided. With a leader who simultaneously encourages expressions of disagreement and works to maintain congenial interpersonal relations among participants, primary group identity need not be associated with quick concurrence and the adoption of a dominant solution. Assuming that disagreements then do indeed surface, additional options may be considered. Under these conditions integrative solutions (like compromises, trade-offs, innovative options) seem possible. Precisely because members share a loyalty to each other, they may search more vigorously for solutions acceptable to all rather than accept one that is opposed by some.

It should be apparent also that these two variables, group norms surrounding how to deal with disagreement and leaders' preferred discussion style, can work in the opposite direction. In effect, there may be group norms that discourage expression of dissent (what Janis expected would happen). Similarly, the group leader may act to suppress any disagreement with the proposed solution. Such leadership will be particularly effective in minimizing disagreement if such an authoritative figure expresses his or her preference for the tabled solution early on and vigorously advocates it.

In essence, group norms regarding how to handle disagreement and leaders' preferred discussion style are two important mediating variables in the Concurrence Model. Depending on their configuration, they either reinforce or modify the effects of groupthink on any group whose members assign it their primary identity. Figure 1 suggests the conditions in the Concurrence Model that result in the adoption of the dominant solution. We can state these expectations as formal hypotheses:

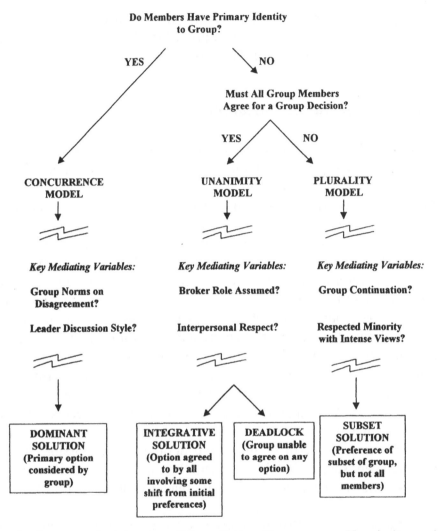

FIGURE 1. Summary decision tree differentiating Concurrence, Unanimity, and Plurality Group Models and the key mediating variables that can reinforce or deflect the most likely outcomes (shown in boxes)

1. Other things being equal, groups whose members have their primary identity with that group are more likely than groups whose primary identity is outside the group to choose dominant solutions to problems.
2. Groups whose members have their primary identity with that group and who have leaders who actively encourage generation of multiple options and their critical appraisal are more likely to choose integrative solutions to problems, whereas groups whose leaders suppress the articulation of multiple options and discourage critical assessment of favored proposals are very likely to choose dominant solutions.

3. Groups whose members have their primary identity with that group and have norms that permit internal disagreement and appraisal of each other's proposals are less likely to choose dominant solutions and more likely to select integrative solutions than such groups with norms that discourage expression of differences.
4. Groups whose members have their primary identity with the group and have both leaders who encourage consideration of multiple options and norms that permit dissent are most likely to choose integrative solutions when compared to such groups that have only either appropriate leaders or norms.

## The 1938 Munich Crisis [8]

British cabinets usually conform to our definition of a single-group decision unit. In August 1938 the cabinet of British Prime Minister Neville Chamberlain received alarming intelligence reports that Germany planned to invade the Sudeten area of Czechoslovakia by October. As the situation unfolded, the cabinet convened repeatedly to decide how to deal with the problem. One of the first such occasions for decision occurred on August 30 in a session attended by eighteen of the twenty-one cabinet members. The foreign secretary, Lord Halifax, opened the meeting with an analysis of the situation. He then recommended that the British government deliberately create ambiguity for Hitler by not indicating whether it would take military action if Germany attacked Czechoslovakia. In effect, Halifax's proposal sought to complicate Hitler's strategy by "keeping Germany guessing." The prime minister endorsed his foreign secretary's proposal. During the ensuing discussion, only two officials spoke unequivocally in favor of another option—to explicitly warn Germany not to attack. At the conclusion of the meeting, Foreign Secretary Halifax responded to those who had urged that a warning be issued followed by the prime minister declaring the cabinet unanimous in support of a policy of deliberate ambiguity with the intention of keeping Germany guessing. No one present objected.

To fit this case and others to be discussed below to the models, Figure 2 presents a more comprehensive decision tree depicting the hypothesized effects of each variable in relation to others. To apply the decision tree in Figure 2 to a specific decision, the analyst must answer a series of questions about the status of key variables in the historical case. Answering each question establishes a dichotomous value for one of the model's variables, that is, it indicates whether the variable was present. The hypothesized relationships among variables in each model appear as the pathways through the decision tree. Figure 2 may

---

[8] This section and other discussions in this article of the 1938 Munich crisis draw from Walker and Watson (1989) and Walker (1990, 1995).

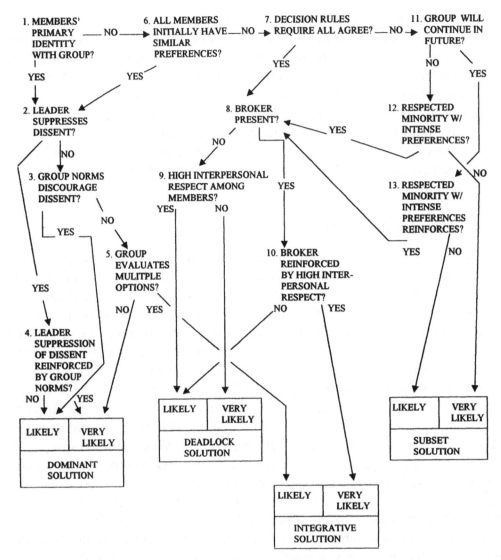

FIGURE 2. A decision tree representation of the effects of key mediating variables on group decision outputs in the three models

seem somewhat complex because it incorporates all three models. As we shall see later, by presenting it in this fashion we can denote how certain mediating variables are able to change the type of decision most clearly associated with one model to that of another.

Consider now the Chamberlain cabinet decision of August 30, 1938, using Figure 2. The first question with which every decision analysis must begin (Members' Primary Identity with Group?) appears in the upper left-hand corner of Figure 2. The major variables associated with the Concurrence Model can be found below that initial question on the left side of the figure. With respect to the initial question, members of the Chamberlain cabinet can be said to have had their primary loyalty to that group. All were members of the same political party and served at the pleasure of the prime minister. Thus, the answer to Question 1 is "yes." This leads us to Question 2: Does Leader Suppress Dissent? Chamberlain clearly served as the group's leader with Halifax as his strong second. Both the prime minister and his foreign secretary made their own preferences clear at the outset, which appears to have had a somewhat chilling effect on the expression of disagreement. Question 2 also can be answered "yes." In Question 4, to which our previous affirmative answer directs us, the variable concerns group norms that reinforce the leader's suppression of dissent. In the British cabinet there is a tradition of permitting ministers to raise questions as well as express their preferences and the record shows that several members did precisely that during the meeting. The answer to Question 4 being "no," the likely outcome expected by the model is the Dominant Solution, that is, concurrence around the option that had been given initial prominence in the group's deliberation. For this particular occasion for decision, the "keep Germany guessing" position initially posed by Halifax, and immediately endorsed by Chamberlain, can be viewed as conforming to the model's expectation. Notice that the model would have expected the dominant solution to have been even more probable (very likely) had group norms reinforced the leader's tendency to discourage dissent.

## UNANIMITY MODEL

### *Resolving Conflict*

The opposite of a group committed to avoiding internal conflict is one that acknowledges disagreement as a fundamental, often unavoidable, part of the policymaking process. People have different information, experiences, and values; therefore, disagreements are natural. As has been suggested, group conflict can be constructive by forcing people to think through their arguments more carefully and, on occasion, to modify their views to take into account the insights and preferences of others. Even if outspoken advocates of competing positions cannot recognize the need and possibility of bridging their differences, other members of a group may be able to do so. From this perspective, disagreement and its resolution become central to the group process.

Recognizing the existence of substantive disagreements among group members is at the core of the theoretical orientation known as bureaucratic politics. Because groups in government are embedded in agencies, ministries, departments, or other bureaucracies, conflicts based on competing missions, interests, and interpretations are structurally built into the interagency process. The formation of policy and other problem-solving tasks frequently involves the attempted reconciliation among these different organizational interests. The entity often faced with resolving such differences is an interagency committee, task force, or council.

Ever since his analytical study of the 1962 Cuban missile crisis, in which bureaucratic politics served as one of three alternative interpretations for explaining American decisions, Allison (1971) has been associated with this approach. And, certainly, Allison assumed a pivotal role in the advocacy of the bureaucratic politics approach and his treatment has been the reference point for numerous applications and evaluations. Unlike the concept of groupthink, however, bureaucratic politics must be seen as the product of many analysts.[9] Distilling the essence of the bureaucratic politics orientation from the multiple developers is a matter of interpretation.

## *Outside Identity and Total Agreement*

One essential feature of bureaucratic politics contrasts directly with notions about groupthink and the Concurrence Model we drew from it. Whereas the Concurrence Model necessitates that group members give their primary identity to the authoritative decision unit (the group), bureaucratic politics expects members of the group to have their primary loyalty elsewhere. They are, in fact, in the group to represent their bureau, agency, or some other outside interest. It is their home bureau whose views they are to present and defend. In all likelihood it is that bureau that pays their salary, determines their promotion, and confers status on them. If a civil servant or political appointee is not supportive of their home agency, that organization will surely look for ways to punish the wayward individual. For the Unanimity Model, we borrow the idea from bureaucratic politics that in certain problem-solving groups the members' primary identity with respect to the issue under consideration lies *outside* the decision group. For our purposes, however, it is not essential that group members represent government agencies; what we are concerned about here is that each group member assigns a higher priority to furthering the interests of some outside entity than those of the present group. This outside identity reduces their concerns with group harmony and makes likely their willingness to express disagreement when the preferences of that outside entity are not being served.

Another characteristic present in some interpretations of bureaucratic politics is also incorporated into the Unanimity Model: the requirement that all members must agree to a position before it can become the group's decision. This inclusiveness condition does not necessitate that all members have equal

---

[9] There have been numerous thoughtful critiques of the Bureaucratic Politics Model including Art (1973), Bendor and Hammond (1992), and 't Hart and Rosenthal (1998).

power. There may be substantial differentials among the members along several dimensions (e.g., status, expertise, and seniority), but no single individual or subset of members has the ability to decide alone. In the end all members must support the decision which means that each can veto a possible proposal.

Other studies of bureaucratic politics offer additional features, but it is these two properties (outside identity and agreement of all) that are incorporated into the Unanimity Model. These properties often lead to a distinctive group process. Because the members' primary identities are outside, group disagreement on substantive issues is quite common. As a result of the fact that each member can veto anyone else's proposal, members must listen and take into account the views of everyone else. In such circumstances, it is not an effective strategy to seek out only those members of the group whose views are similar to one's own and construct a faction with them because the excluded members can block the resulting option.

Do groups fulfilling the Unanimity Model conditions actually exist in foreign policy situations or other policymaking circumstances? In an earlier time, the American Joint Chiefs of Staff exemplified this model. Today many juries do. Gaenslen (1996) found examples—including the Brezhnev Politburo—in Russia, China, and Japan as well as the United States. If one acknowledges that major foreign policy matters are typically the function of a country's chief executive, then the expectation might be that this head of state is always in a position to be first among equals, if not the outright authoritative policymaker. We would like to argue here that a subordinate interagency group, however, may keep their decision away from a higher authority fearing that it will lose control of the problem, face an outcome it does not want, or be judged as incompetent. Leaders for their part may insist on a unified recommendation from an advisory group, delegate their authority to the group, or simply lose track of the problem enabling others to make the decision. Moreover, even without a formal rule requiring unanimity, representatives of agencies that need to work together on a range of issues may come to understand that taking action in one instance contrary to the preferences of one of their number may be shortsighted in the long run. In the future, the omitted agency may be in a position to ride roughshod over the preferences of its prior antagonists. An understanding of this feature of bureaucratic life creates the conditions under which groups may come to operate with a consensus or unanimity requirement.

What are the likely decision outputs associated with the Unanimity Model? Two different results are expected if no intervening mechanisms are introduced into the group dynamic. One outcome is deadlock. Group members are simply unable to bridge their differences and stalemate results. The other outcome is the integrative solution. Integrative solutions may constitute a compromise among members in which they accept a solution that is of lower value to them than their first choice. Alternatively, a solution may be discovered by the group in the course of their deliberations that fulfills the interests of everyone. Bureaucratic politics scholars frequently mention a third type of integrative output that involves a trade-off among members or side payments to some holdouts to secure their support which is for the moment not a part of the present model.

## *Unanimity Mediating Variables*

At this point, let us introduce two characteristics we believe differentiate the likelihood of deadlock from achieving an integrative solution. One such mediating variable is the presence of one or more members who assume a "broker role." The other is the degree of interpersonal respect among members.

The term "broker role" is used to identify an explicit set of activities that one or more individuals in the group assume who seek to find a solution to the problem that addresses the concerns and interests of all members. It is a particular type of facilitation that involves clarifying each member's concerns, proposing solutions that attempt to bridge differences, and encouraging others to invent such options. To be classified as playing a broker role, an individual must make repeated and varied initiatives of this kind rather than engage in a singular intervention in the group process. It is possible that more than one person in a group will assume such a role. Moreover, it need not be the formal chair or leader. The presence of a broker provides no assurance of an integrative solution in the group, but such participation increases the likelihood of such a decision and reduces the chances of deadlock.

The degree of interpersonal respect members have for one another is the other mediating variable in the Unanimity Model. In the previously described Concurrence Model, we expected members to have a high degree of interpersonal respect for one another because of their primary identity with that group. No such condition is necessary in the Unanimity Model whose members have their primary identity elsewhere. In fact, in such groups it is easy to imagine that members may regard others in the group as the opposition and consider these others' positions and arguments as unreasonable, foolish, or deliberately intended to damage their own well-being. If the members lack respect for at least some other members of the group, it is unlikely that they will devote much effort to finding a solution that incorporates the other members' concerns. Their energy will more likely be devoted to finding ways to get them out of the group or creating an alternative decision unit that excludes them. Although such low levels of interpersonal respect may exist between some members of Unanimity Model groups, it is a variable and not a constant. Thus, when members of such groups have positive views of one another, even though they disagree on the means of coping with the immediate problem, they are more likely to be prepared to work together for a mutually acceptable resolution. In such cases, the likelihood of an integrative solution should increase with a corresponding decrease in the occurrence of deadlock.

The expectations associated with the Unanimity Model, shown in Figure 1, are captured in the following formal propositions:

1. Other things being equal, in groups whose members have their primary identity elsewhere but must all agree to any proposal for dealing with the problem, deadlocks and integrative solutions are the most likely outputs.
2. In groups whose members have their primary identity elsewhere but must all agree to any proposal for dealing with the problem, when one or more members assumes a broker role, integrative solutions become more likely and deadlock less likely.
3. In groups whose members have their primary identity elsewhere but must all agree to any proposal for dealing with the problem, the higher the degree of interpersonal respect among group members, the more likely are integrative solutions and the less likely is deadlock.
4. In groups whose members have their primary identity elsewhere but must all agree to any proposal for dealing with the problem, the presence of both the broker role and a high degree of interpersonal respect increase the likelihood of integrative solutions more than in such groups that have only the broker role *or* a high degree of interpersonal respect.

## *Soviet Submarine Aground in Sweden, 1981* [10]

Sweden has long practiced a foreign policy of armed neutrality. Throughout the Cold War it used its diplomacy and military to ensure that both superpower blocs respected its position. Thus, the Swedish government confronted an urgent problem when in the morning of October 28, 1981, they discovered a Soviet submarine trapped on the rocks in a restricted military coastal area. Initial Swedish decisions involved a set of demands to the Soviet government that the latter appeared ready to accept. Then the problem became more complicated. From outside the hull, Swedish defense experts detected radiation that led them to conclude nuclear weapons might be present.

For Sweden, the nuclear possibility changed the definition of the problem. Not only had a foreign submarine violated restricted coastal territory, but there was also an indication that it violated Swedish insistence on keeping nuclear weapons out of its territory. If the Soviets brought nuclear weapons into Swedish territory, sterner measures might be required. The Swedish government

---

[10] This section and other discussions in this article concerning the Soviet submarine that ran aground in Sweden in 1981 draw upon Sundelius (1990) and Stern and Sundelius (1992a, 1992b).

requested that its representatives be allowed on board to inspect the submarine's torpedo tubes to make a definitive determination. The Soviets refused; instead, they responded by building up their naval forces just outside the restricted territorial waters. Now the Swedish government had to decide its next move.

The Soviet ambassador conveyed his government's rejection late on Wednesday afternoon, November 4, 1981. That evening the Swedish prime minister convened an ad hoc group including, among others, his foreign minister, the permanent under secretary of foreign affairs, various other diplomatic and military officers, as well as the leaders of the two major opposition parties in the Parliament. Those present disagreed over whether to hold the Soviet submarine until an inspection was permitted or to ignore the possibility of nuclear weapons and tow the sub to open waters. Given the strong norms in Swedish policy for forging consensus, a compromise emerged. The policymakers agreed to a strong diplomatic protest note and to condemn publicly the Soviet Union for entering Swedish waters with a ship that likely carried nuclear weapons. Simultaneously with the protests, the Swedes would tow the submarine to international waters without an onboard inspection. Their solution met the requirement of some members of the group not to ignore the nuclear weapons issue while, at the same time, fulfilled the preference of other group members not to escalate the crisis dangerously.

To apply this case to the model, we again turn to Figure 2. The initial question concerning whether members' primary identity is with the group is answered in the negative. Recall that the ad hoc group consisted of both representatives of the foreign ministry and military and the two leaders of the major opposition parties. Moreover, members initially had different preferences on how to treat the problem (Question 6). Some wanted to confront the nuclear weapons issue and argued to hold the submarine until the Soviets agreed to an inspection. Others argued against a serious escalation of the crisis with a super power. Swedish requirements for consensus made unanimity the necessary decision rule (Question 7). Indeed, unanimity is an integral part of the Swedish foreign policymaking process; there is a normative aspiration to reach a common position based on shared values (Sundelius, 1989). At this point, Question 8 asks that we determine whether a member of the group played a broker role. We have no direct information on that matter, but it can reasonably be inferred that one or more individuals sought to devise a compromise. Given the pressure in Swedish policymaking for consensus and the evidence that a compromise was reached, the assumption that a broker was present seems plausible. Moreover, given the composition of this ad hoc group to which the prime minister had invited leaders of the political opposition, there would seem to have been a high level of interpersonal respect (Question 10). By tracing this path through the decision tree in Figure 2, we see that the model regards an integrative solution as very likely. The actual compromise involving a public and private protest of a probable nuclear violation combined with the release of the submarine fits that expectation.

## PLURALITY MODEL

At the core of the two previous models were opposing strategies for the management of disagreements in groups coping with a policy problem. In the Concurrence Model, conflict is avoided by all members as they promptly accept the initially advanced proposal. In the Unanimity Model, conflict is resolved by working through the disagreement until all members agree on a particular solution. The Plurality Model represents a third perspective on substantive conflict in groups. It acknowledges that disagreements are likely, but that achieving a resolution acceptable to all members may not be possible or may be achieved only at considerable costs—in the time required or in the creation of an awkward compromise satisfactory to no one. Some group decision, however, may be essential. Solutions that do not further the objectives of at least some substantial portion of the group members may be difficult to implement and enforce.

To avoid these potential difficulties, the Plurality Model recognizes that not all disagreements may be capable of resolution to the satisfaction of all members. In this perspective, a group can reach a decision when a certain proportion of the members concur, even though others may still remain opposed. In brief, in the Plurality Model members agree a priori to comply with a rule that permits acceptance of a solution when a certain proportion of the members agree. Although a plurality decision rule need not require agreement among more than half the members, majority rule (e.g., 50 percent plus 1; 2/3; 3/4, etc.) has been the most frequently studied form.[11]

In democratic political theory, the relationship between majorities and minorities—the divisions created by majority rule—has long been a topic of great concern (e.g., Chapman and Wertheimer, 1990). Comparing majority vs. unanimous decisions has engaged students of rational choice (e.g., Buchanan and Tullock, 1962; Rae, 1975). In addition, some American states and other jurisdictions have begun to introduce the practice of permitting a jury to reach a decision if two thirds or three fourths of the panel agree, rather than following the previously required unanimous verdict. Legal scholars have asked: Does it make a difference in decisions of guilt or innocence if all members must agree on the verdict or only a majority? In the United States, the matter has come before the Supreme Court (e.g., *Johnson vs. Louisiana* and *Apodace, Cooper,*

---

[11] Under certain conditions, a group may accept a plurality rather than a majority rule or they may agree that certain members must be included in the winning subset. Granting some but not all members a veto power has this effect.

*and Madden vs. Oregon*). In part prompted by the legal interest, a body of research has appeared in political science and social psychology. The findings from this literature have been reviewed by Miller (1989).

A broad array of studies have confirmed the expectation that groups governed by a majority rule typically take less time to reach a decision than those functioning under the requirement of unanimity. Those familiar with the public policy arena know that time is often critical, making a less stringent decision rule attractive. Beyond the time variation, however, there appear to be different dynamics within groups using some form of majority rule rather than unanimity. Under a unanimity rule, participants recognize that no solution is possible unless it is accepted by everyone. This recognition requires each person to take into account the views of all other members. Efforts at persuasion and the search for consensus must be addressed to all. In contrast, in a majority rule situation, persons can concentrate attention on those members whose views seem most similar to their own (or on persons who appear to be undecided or most open to persuasion). Those with extremely different views can be ignored in building a winning majority position.[12] This rationale provides the basis for our expectation that, other things being equal, groups whose members have their primary identity elsewhere will most frequently reach a solution that favors the position of a subset of their members.

## *Plurality Mediating Variables*

As with the other models, the presence of certain mediating variables can alter the likelihood of groups representing the Plurality Model reaching a subset solution. Two that seem of particular importance are the distribution and intensity of preferences among members and the continuity of the group. Distribution of preferences refers to the degree of similarity or proximity among members' positions and interests and intensity refers to how strongly members value their preferences. For the necessary majority to form, a sufficient number of members must have or invent an option that is acceptable to them but not necessarily to all members. If the preferences and underlying interests of no set of members are more proximate than those of any others in the group or if those with somewhat similar views strongly resist concession, then a stable majority is unlikely. In such a case, either an option can be formed that is as acceptable to all members as it is to any potential subset and the decision output is an integrative solution, or the group deadlocks.

---

[12] Riker's (1962) classic study of coalition formation, which triggered considerable research, argued that under many conditions groups would seek to include only the minimum number of members necessary for a winning coalition in order that any gains from success could be maximized among the winners.

Several scholars have examined the duration, continuity, and life span of decision groups and the effects of a group's current life stage on problem solving. For example, Stern (1997) explores what he calls "new group syndrome," or the consequences that emerge from groups that have only recently formed. Moreland and Levine (1988) have conducted research on the entire life cycle of groups. For our purposes, it is the very continuation of the group that is at issue. Do members expect the group to disband after it has dealt with the current problem or do they anticipate that it will continue for some time and deal with a range of other problems in the future? Knowing that the group will subsequently address other problems with the same members, a majority subgroup may be more reluctant to select an option that one or more members in the minority intensely oppose. In the future, those in the current majority may find themselves in the minority position on some issue that is of great concern to them. Thus, when the members expect the group to continue and there is a current minority with intensely held preferences, a greater effort will be made to find a solution that addresses some of the minorities' concerns. In such cases, an integrative solution is more likely even under majority rule. Conversely, when the group members do not expect to work together in the future on other issues, no such constraint is probable and a subset solution can be expected.

The right side of Figure 1 depicts the decision output most generally expected in the Plurality Model (a subset solution) and identifies key mediating variables. Examination of this path in the figure may suggest to the reader the existence of a tautology. If the definition of the Plurality Model includes the stipulation that all group members need not agree for a decision to be reached in that group, then is it not true by definition that the outcome will be a subset solution (i.e., a decision representing the preferences of some portion of the members but not all)? The subset solution is certainly expected when all members need not agree, but it is not *required* that such a group in disagreement must reach a decision that excludes some members. What defines a Plurality Model group is its operation under a rule that *permits* decision without complete agreement. Moreover, as indicated in the hypotheses below, some mediating variables are expected to influence the relationship between the group's decision rule and the expected outcome. The anticipated relationships for the Plurality Model are summarized in the following hypotheses:

1. Other things being equal, groups whose members have their primary loyalty elsewhere and are not required to achieve unanimity for agreement are very likely to reach subset solutions when members disagree.
2. In groups whose members have their primary loyalty elsewhere and are not required to achieve unanimity for agreement, if the minority in a dispute includes respected members who hold their preferences intensely, then deadlock or integrative solutions are more likely than subset solu-

tions, and, conversely, if none in the minority are held in high esteem or do not intensely advocate their preferences, a subset solution is increasingly likely.
3. In groups whose members have their primary loyalty elsewhere and are not required to achieve unanimity for agreement, if the group is expected to address other problems in the future and a current minority hold their preferences intensely, then an integrative solution is more likely, and, conversely, if the group is not expected to address future issues or if the minority does not express intense preferences, a subset solution is increasingly likely.
4. In groups whose members have their primary loyalty elsewhere and are not required to achieve unanimity for agreement, if neither a respected minority with intense preferences nor an expectation that the group will continue and address other future problems is present, then a subset solution is more likely than if only one of these two conditions is present.

## Israeli Response to Egyptian Military Actions, May 1967 [13]

The Plurality Model can be illustrated by a study of the Israeli decisions in response to threats from Egypt and other Arab states in the spring of 1967. A critical development in what was perceived by all as a deteriorating situation occurred late on May 16 when Egyptian President Nasser demanded the withdrawal of U.N. peacekeeping forces from the Sinai and subsequently moved Egyptian military forces into that desert bordering Israel. Then on May 22, Egypt blockaded the Straits of Tiran, preventing shipping access for Israel through the straits. This action triggered an urgent meeting of Israeli policymakers on May 23. They met first as the Ministerial Committee on Defense and then as the cabinet. These policymakers had to deal not only with the military actions of their Arab neighbor, but also with a request from United States President Lyndon Johnson who urged Israel to defer any action for forty-eight hours to allow the Americans to explore possible international response. Military Chief of Staff Rabin argued that to maintain the credibility of Israel's deterrence policy immediate action was required. Foreign Minister Eban countered that any action must be deferred until consultation with the United States had occurred. During the ensuing deliberations, Rabin and the military leadership acknowledged that a delay of forty-eight hours in the initiation of military action would not be critical. Subsequently the cabinet unanimously agreed to the Eban position of consultation.

---

[13] This section and other discussions in this article of Israeli decisions regarding Egyptian military actions in May 1967 draw from Stein and Tanter (1980) and Stein (1990, 1991).

Because the Israeli cabinet operates with a majority voting rule, the Plurality Model clearly applies. More specifically, we begin our analysis in Figure 2 by establishing that the primary loyalty of members of the Ministerial Committee on Defense and the cabinet lies outside those groups (Question 1). In the spring of 1967, the Israeli cabinet included representatives from several political parties; it conforms to our definition of a group (rather than being considered a coalition) because in this crisis, cabinet members acted more or less as free agents instead of representatives who were expected to consult with their outside constituents before taking, or changing, their positions. Did cabinet members primarily identify with the cabinet, that is, the decision-making group? Given that the Israeli policymakers perceived themselves facing a grave national security crisis, one could argue that most group members gave their primary loyalty to the group. Although such an interpretation is plausible, it will be proposed here that the members' loyalty was to the nation in the crisis, but not necessarily to the cabinet. The group consisted of members of multiple political parties and of professional military and diplomatic personnel. It seems reasonable to assume that these individuals had their longer-term allegiance to these other entities rather than to the cabinet as a body. Thus, Question 1 in Figure 2 requires a negative answer. Unambiguously the answer to the next query along the path is "no" (Question 6: Do all members initially have similar preferences?). During much of its deliberation, cabinet members were divided between the opposing views of Foreign Minister Abba Eban and Chief of Staff Yitzhak Rabin. This leads to Question 7 that asks whether all members must agree for there to be a decision. Again, the answer is "no." The cabinet, however, is a continuing body that will deliberate on other issues; therefore, Question 11 must be answered "yes." At Question 13, we must decide whether respected minorities in the group have intense preferences that reinforce the mediating variable of a continuous group. That Rabin was a highly respected member of the cabinet is clear, but did he and his colleagues hold "intense" preferences for immediate action? As we observed earlier, under extensive questioning, the military experts acknowledged that a delay of action for forty-eight hours would have little consequence; when a final vote was taken the option of deferring action was accepted by all. Since Question 13 requires that the respected minority have intense preferences, it must be answered in the negative in this case. That response ends the path through the decision tree. Under the conditions specified, the expectation from the model is that a subset solution is likely. Does the actual Israeli cabinet decision in this case represent a subset solution? In the final vote, all members agreed to a deferral. The definition of a subset outcome recognizes the possibility of a shift in some members' preferences, but still assumes the final choice does not represent the preferences of all. In the end, did the actual preferences of Rabin and the other initial advocates of immediate action change? There is some evidence that those favoring Rabin's position believed they still had time to take action and that the deferral would not limit their future options either through the allocation of resources or through the generation of expectations in others. It was still early in the process. They had not changed their original preferences but were willing to secure additional information before choosing to attack.

## FURTHER OCCASIONS FOR DECISION FROM THE HISTORICAL CASES

At the outset of this essay, we suggested that decisions corresponding to the Concurrence Model are likely to involve dominant solutions. The dynamics within Prime Minister Chamberlain's cabinet illustrated how such a decision could occur. In the Unanimity Model, decisions tend toward either deadlock or an integrative solution. The Swedish response to possible nuclear weapons on the grounded Soviet submarine provided an example of a group—that required consensus—finding an integrative decision. Finally, groups operating in accordance with the Plurality Model are hypothesized to reach solutions preferred by a subset of the members. The Israeli decision in 1967 arguably indicates that a subset of the members of the cabinet got their preference adopted and that even though Rabin continued to want immediate action, he elected not to voice an objection to the majority preference given the belief there was still time left in which to act.

It is a major contention of this piece, though, that key mediating variables can alter the expected decision outcomes for each of the three models. A mediating variable either reinforces the model's disposition toward a certain outcome or alters it. Figure 2 reveals a variety of hypothesized paths in which the value of one or more mediating variables shifts the likely outcome from that associated with one model to that of another. In other words, the outcomes are contingent upon the status of relevant mediating variables.

To demonstrate this contingency feature, it is instructive to return to the three historical cases. The policymakers in each case were involved in a series of decisions in their efforts to cope with the problem at hand. The three previously reviewed decisions each isolate a single occasion for decision from what were, in fact, a string of occasions for decision that faced each set of policymakers across a period of time. In these other occasions for decision, the status of certain of the mediating variables changed from what they were in the decisions described above. Under these changed conditions, the models lead to different expected decision outcomes.

Recall that when Prime Minister Chamberlain's cabinet met in August 1938 to consider what to do about reports from British intelligence that Hitler was planning to invade Czechoslovakia, Chamberlain and his foreign minister acted as a united team. Foreign Minister Halifax proposed a policy at the outset of the meeting; Chamberlain immediately endorsed it. In the ensuing cabinet discussion there was little dissent and Chamberlain declared that the recommendation proposed by Halifax was adopted.

In the following weeks, Prime Minister Chamberlain met twice with Hitler. After the second meeting, Chamberlain returned with written demands from Hitler. The British cabinet met four times on September 24-25, 1938, to decide on a response. In the first meeting, Hitler's memorandum was distributed and Chamberlain proposed Hitler's terms be accepted. The meeting then adjourned until the next day. At the second meeting, Foreign Minister Halifax started the meeting by recommending that Hitler's memorandum be rejected.

From the perspective of the Concurrence Model, a key difference emerges when these two occasions for decision are compared. Whereas earlier Chamberlain and Halifax had acted jointly to suppress disagreement, now Halifax begins by opposing his prime minister's recommendation. Under this latter condition, Chamberlain makes no serious attempt to curb disagreement among his cabinet. Thus, the mediating variable concerning the leader's effort to curb group dissent is different. In Figure 2, the answer to Question 2 shifts from "yes" (in the earlier decision) to "no." This change calls for exploring a different pathway through the decision tree. As we previously determined, the norms of the British cabinet do not discourage the expression of dissent (Question 3). In the second cabinet meeting held on September 25, multiple options are discussed (Question 5). This described path through the decision tree leads to the expectation that an integrative solution is likely.

The historical case appears to conform to this expectation. In the second cabinet meeting on September 25, the prime minister proposed an alternative approach that involved his writing a final letter to Hitler urging the Nazi leader to negotiate a compromise. Chamberlain's letter was to be delivered personally by a high-level British official. If in that meeting Hitler showed no disposition to compromise, then the British official would deliver a firm warning. At that meeting the cabinet debated Chamberlain's new proposal and certain modifications were made. They then adopted the new proposal.

The consideration of these two different occasions for decision by the Chamberlain cabinet underscores that a change in certain key process variables can alter the likely decision. The same group cannot be assumed to conform to a constant model of decision making even when they are considering the same problem. This point also is confirmed by the Swedish decisions on the Soviet submarine.

In the earlier account of the Swedish management of the Soviet submarine that had run aground on their territory, we examined their decision making after officials detected radiation at external points along the hull of the ship. Days earlier, the government had worked through an initial occasion for decision that arose when they first learned (on October 28, 1981) that a Soviet submarine was aground in Swedish waters. The earlier, critical decision appears to have been made by an ad hoc group meeting in the Foreign Ministry on the morning of October 29. The permanent under secretary of foreign affairs presided over a group composed of several of his diplomatic colleagues and their counterparts in the Ministry of Defense as well as the supreme commander of the armed forces. The group's deliberations began with a briefing by the Foreign Ministry's international law expert who recommended a course of action and presented such a convincing case for his preference that other avenues were not fully explored. Considering the position of this expert in the Foreign Ministry and the likelihood that he had had discussions before this meeting with the permanent under secretary, it is highly probable that the articulated option was not markedly different from the latter's views. At the conclusion of the meeting, the chair of the group from the Foreign Ministry, his counterpart from the Ministry of Defense, and the military supreme commander briefed the prime minister and foreign minister on the group's recommendation which the political leaders accepted and ordered implemented.

Without reconstructing the detailed pathway in Figure 2 that this decision process appears to have followed, it seems evident that what started out as a candidate for the Unanimity Model—similar to the previously considered Swedish decision—got sidetracked. We find a recognized expert advocating a position at the outset of the deliberations. His position appears to have been endorsed by the group's chair. These similar initial preferences and their likely creation of an atmosphere not conducive to dissent become key switching variables. We have been told that the Swedish decision norm is for consensus (unanimity). Yet, in response to this occasion for decision, the actual decision is a dominant solution, as the decision tree path in Figure 2 would lead us to expect.

No less variation occurs between occasions for decision in the Israeli government's crisis of 1967. The Israeli decision—described previously—to defer an action for at least forty-eight hours had occurred on May 23. As a result, Foreign Minister Eban left to consult in Paris, London, and Washington. While he was away, the situation deteriorated. Egyptian troops continued to move into forward positions and on May 25 Israeli intelligence intercepted an order to Egyptian air force units instructing them to prepare to attack two days later. Senior Israeli military officers now urged an immediate preemptive attack. Meanwhile, U.S. President Lyndon Johnson urged a British plan for an international naval flotilla to open the blockaded strait. The Israeli cabinet began what became an all-night deliberation that continued the next day. In an informal vote early on the morning of May 28th, the cabinet was deadlocked as ministers divided evenly in favor of and against attack. At that point, Prime Minister Eshkol recessed the session for a few hours of rest. During the break in the cabinet meeting, messages arrived from President Johnson and Secretary of State Dean Rusk giving an encouraging assessment of the potential for an international flotilla and warning Israel against unilateral action. When the cabinet reconvened, the prime minister and several others shifted their vote to support delay.

This decision sequence poses a fascinatingly intricate pathway cutting across mediating variables in the decision tree of Figure 2. Primary loyalty of members lies outside the group (Question 1) because of the cabinet's multiparty composition. Of course, in this case a clear disagreement existed from the outset (Question 6) in this group with established procedures that permit decisions by a majority vote (Question 7). The cabinet is a continuing group (Question 11). The prime minister and his foreign minister, both respected leaders, were among those on opposing sides who felt so strongly about their positions that an all-night session failed to change any views (Question 13). Inasmuch as all cabinet members appeared to have taken positions, no one appears to have been available to play a broker role (Question 8). Given those who were on the opposing sides, it can reasonably be assumed that there was substantial interpersonal respect among the members (Question 9). This protracted path through Figure 2 leads to a deadlock as the likely outcome. And when Prime Minister Eshkol called a recess in the morning that was the actual situation.

When, however, the cabinet reconvened in the afternoon, members had new information in the form of communications from the U.S. president and his secretary of state. Based on those communiques, Eshkol and the other members of his party in the cabinet changed their positions to favor delay; one member abstained. When a new vote was taken, only the minister of transportation, Moshe Carmel, opposed deferral of action; he was not a member of the ministerial committee on defense, the cabinet subgroup with special responsibility for security matters.

It is precisely the status of this individual who remains in opposition that must be assessed in applying the models in Figure 2. In this new Israeli occasion for decision following the messages from the United States, the answers to Questions 1, 6, 7, and 11 remain as before. What appears to change is the response to Question 13 that asks if there is also a respected minority with intense preferences. There can be little doubt of the intensity of Minister Carmel's views given that in the end he alone voted against delay. If, however, we infer that on issues of national security the minister of transportation's views may not be respected by others as reflecting considerable expertise or authority, then the outcome expected by the model is a solution representing the preferences of a subset of the group members. On the afternoon of May 28, 1967, the Israeli cabinet decided that way.

## Conclusions

The examination of the additional occasions for decision in the historical cases illustrates an important point. Some of the earlier explanations of group decision making in the literature have tended to advance unqualified interpretations. Yet, as we have seen, the presence and status of certain mediating variables can dramatically shift the likely decision outcome. To oversimplify slightly, a situation that bears some of the hallmarks of groupthink may, in fact, generate a decision that one might more readily associate with bureaucratic politics. Thus, an important task is to identify the critical intervening group structure and process variables that play a role in shaping the outcome. In short, the models we have presented here regarding the effects that different conflict management strategies can have on group decision making must be viewed as contingent models.

Some mediating variables seem more likely than others to change values from one decision to the next. In general, one might expect process variables (e.g., a member's willingness to assume a broker role) to vary more readily than structural variables (e.g., decision rules or group norms). Yet even some of these seemingly more enduring group properties can change abruptly if the politically motivated group must deal with a variety of different kinds of problems.

Our review of multiple occasions for decision in the same historical cases introduces another issue. The models presented in this paper "slice" the decision process into narrow, discrete decisions. The British cabinet's dealings with Hitler in 1938, the Swedish government's response to the Soviet submarine in 1981, and the Israeli decisions about war in 1967, however, demonstrate that foreign policy decision making is seldom a "one-shot" task. Instead, foreign policy problems unfold over a period of time in part affected by policymakers' prior actions. Policymakers frequently engage in sequential decision making in which more or less the same individuals find themselves dealing with the same problem repeatedly with their current decisions being affected by what has gone before. In future theoretical development, models such as those proposed in this essay must be embedded in the longer stream of policymaking.

We must also recognize how the conflict management strategies that groups use fit into this larger context. This essay has argued that for numerous reasons people working in policy groups on problems with no known correct answer are likely to disagree about what is to be done. And such disagreements may extend to differences concerning the nature of the problem they are addressing (i.e., problem definition or framing issues). If these disagreements are permitted to surface, they can create conflict within a group. Groups tend to develop different dispositions toward the management of conflict among their members. We have suggested they can deny (or ignore) disagreement, they can resolve it, or they can accept it. These differences provided the point of departure for this essay. At a macro level, the reasoning is that these dispositions toward conflict lead to group procedures regarding how to come to some interpersonal agreement (Concurrence, Unanimity, or Plurality Models). These procedures, in turn, affect the probable solutions the group will reach (dominant, deadlock, subset, or integrative). Certain group structure and process variables make the expected type of decision output more or less likely.

Let us conclude with a brief consideration of the implications of the different decision outputs associated with these models. Does it make a difference in terms of the quality of the decision if the group dynamics involve concurrence, unanimity, or plurality? Discussion of such a question must always begin with the caveat that process cannot ensure the quality of results. As has been noted before, good decisions can flow from a flawed process and fiascoes can emerge from a quality process. We would expect, however, that sound decision processes, rather than poor ones, would increase the chances of decision outputs that do what the policymakers intended.

Janis (1972, 1982) makes a strong case for the greater possibility of flawed decision making when a single option is accepted without critical evaluation or thoughtful comparison to other potential solutions. The critiques of Janis and the groupthink perspective generally have not disagreed with the potential harm of premature closure on a group solution.[14] As we observed earlier, concerns with such a flawed process led George (1972) to propose multiple advocacy. In this essay, the process associated with the Concurrence Model and a dominant solution carry the same risks for reaching a quality decision.

The merits of the processes associated with majority vs. unanimity decision rules are more varied. In majority rule vs. unanimous rule juries, Hastie and his associates (1983) distinguished between verdict-driven (majority) and evidence-driven (unanimous) juries, or what Gaenslen (1996:32) has subsequently characterized as "more data-driven than solution-driven, with unanimity-seekers more likely to ask 'what has happened and why?' and majority-seekers more likely to ask 'what should we do?'" With a majority orientation, individuals focus on finding the needed number of people in the group who agree with them. When unanimous agreement is required, it is necessary to engage everyone in the group concerning their reasoning and the information they have in order to reach a decision. This deeper and more extended exchange in groups

---

[14] Critiques of groupthink have been extensive. They have questioned both the antecedents and some of the effects that Janis attributed to premature closure. They have also challenged whether some of the foreign policy episodes that he used as illustrations resulted from the decision process. Despite this array of concerns, it has never been suggested that the uncritical acceptance of one solution is a sound decision process.

with a unanimity rule can lead to divergent thinking and the generation of solutions different from those initially envisioned by anyone.

In an experimental study in which groups worked to determine the right answer to math problems, those that got the correct answer usually had one or more members who determined the right answer and then persuaded the others. The authors, however, reported an unexpected finding. "The emergence of a correct solution in groups containing no correct members is a finding not previously reported. This 'emergence of truth' occurred primarily in groups assigned a unanimous consensus rule" (Stasson et al., 1991:32).

Although the problem-solving groups of present interest are neither juries with dichotomous choices nor students solving problems with known right answers, the examples point to a potential strength of decision processes under a unanimity rule. Such groups display a greater tendency to create or invent new solutions to problems. In the attempt to find a solution that works for everyone, unanimity seems to encourage participants to go beyond their initial positions. Such a procedure would appear to have considerable benefits in political situations.

The weaknesses of the unanimity rule when compared to majority rule include both the greater time that usually is involved and the greater tendency toward stalemate or deadlock. In situations such as crises where a prompt decision is required, the liability is clear. There may be additional benefits for majority rule groups that deal with a problem that evolves through time and requires careful monitoring to determine if the initial solution is working. If a majority-rule group takes action even though some of its members oppose the decision, that minority may be more alert to negative feedback.[15] Since they questioned the majority's solution from the outset, they are likely to be more vigilant in monitoring for indications that the prior response was mistaken. Assuming that the minority has not been removed from the decision group, they may be able to alert the group to the need for further consideration based on feedback that the majority has missed. By contrast, some evidence suggests that groups who have worked hard to build unanimous consent are very reluctant to reopen the deliberation and therefore may be slow to react to evidence that their decision is not achieving its purpose.

It is unlikely that there is one decision process that works best for all situations and problems. The implications for quality associated with each, however, add incentives for determining the conditions under which various processes emerge in a group and their impact on decision outputs. We have argued here that how a group handles disagreements may be one important key.

---

[15] See Billings and Hermann (1998) for the development of this role of minorities in sequential decision making.

## REFERENCES

Allison, Graham T. (1971) *Essence of Decision*. Boston: Little, Brown.

Art, Robert J. (1973) Bureaucratic Politics and American Foreign Policy. *Policy Science* **4**:467–490.

Bendor, Jonathan, and Thomas H. Hammond (1992) Rethinking Allison's Models. *American Political Science Review* **86**:301–322.

Billings, Robert S., and Charles F. Hermann (1998) "Problem Identification in Sequential Policy Decision Making." In *Problem Representation in Foreign Policy Decision Making*, edited by Donald A. Sylvan and James F. Voss. New York: Cambridge University Press.

Buchanan, J. M., and G. Tullock (1962) *The Calculus of Consent*. Ann Arbor: University of Michigan Press.

Chapman, J. W., and A. Wertheimer (1990) *Majorities and Minorities*. New York: New York University Press.

Gaenslen, Fritz (1992) "Decision Making in Groups." In *Political Psychology and Foreign Policy*, edited by Eric Singer and Valerie M. Hudson. Boulder, CO: Westview Press.

Gaenslen, Fritz (1996) Motivational Orientation and the Nature of Consensual Decision Processes. *Political Research Quarterly* **49**:27–49.

Garrison, Jean A. (1999) *Games Advisors Play*. College Station: Texas A & M University Press.

George, Alexander L. (1972) The Case for Multiple Advocacy in Making Foreign Policy. *American Political Science Review* **66**:751–785.

George, Alexander L. (1997) "From Groupthink to Contextual Analysis of Policymaking Groups." In *Beyond Groupthink: Political Group Dynamics and Foreign Policymaking*, edited by Paul 't Hart, Eric K. Stern, and Bengt Sundelius. Ann Arbor: University of Michigan Press.

Halperin, Morton H. (1974) *Bureaucratic Politics and Foreign Policy*. Washington, DC: Brookings Institution.

't Hart, Paul, and Uriel Rosenthal (1998) Reappraising Bureaucratic Politics. *Mershon International Studies Review* **42**:233–240.

't Hart, Paul, Eric K. Stern, and Bengt Sundelius (1997) *Beyond Groupthink: Political Group Dynamics and Foreign Policymaking*. Ann Arbor: University of Michigan Press.

Hastie, Reed (1993) *Inside the Juror*. New York: Cambridge University Press.

HASTIE, REED, S. D. PENROD, AND NANCY PENNINGTON (1983) *Inside the Jury*. Cambridge, MA: Harvard University Press.

HUNTINGTON, SAMUEL P. (1960) Strategic Planning and the Political Process. *Foreign Affairs* **28**:285–299.

JANIS, IRVING L. (1972) *Victims of Groupthink*. Boston: Houghton Mifflin.

JANIS, IRVING L. (1982) *Groupthink*, rev. ed. Boston: Houghton Mifflin.

KAMEDA, T., AND S. SUGIMORI (1993) Psychological Entrapment in Group Decision Making. *Journal of Personality and Social Psychology* **65**:282–292.

LONGLEY, J., AND DEAN G. PRUITT (1980) Groupthink: A Critique of Janis' Theory. *Review of Personality and Social Psychology* **1**:74–93.

MCCAULEY, C. (1989) The Nature of Social Influence in Groupthink. *Journal of Personality and Social Psychology* **57**:250–260.

MCGRATH, JOSEPH E. (1984) *Groups: Interaction and Performance*. Englewood Cliffs, NJ: Prentice-Hall.

MILLER, C. E. (1985) Group Decision Making Under Majority and Unanimity Decision Rules. *Social Psychology Quarterly* **48**:51–61.

MILLER, C. E. (1989) "The Social Psychological Effects of Group Decision Rules." In *Psychology of Group Influence*, 2nd ed., edited by P. B. Paulus. Hillsdale, NJ: Lawrence Erlbaum.

MINTZ, ALEX, NEHEMIAH GEVA, S. B. REDD, AND A. CARNES (1997) The Effect of Dynamic and Static Choice Sets on Political Decision Making. *American Political Science Review* **91**:553–566.

MORELAND, RICHARD L., AND JOHN M. LEVINE (1988) "Group Dynamics Over Time: Development and Socialization in Small Groups." In *The Social Psychology of Time*, edited by Joseph E. McGrath. London: Sage.

NIELSON, M. E., AND C. E. MILLER (1997) The Transmission of Norms Regarding Group Decision Rules. *Personality and Social Psychology Bulletin* **23**:516–525.

PRESTON, THOMAS (2001) *The President and His Inner Circle*. New York: Columbia University Press.

RAE, D. (1975) The Limits of Consensual Decision. *American Political Science Review* **69**:1270–1294.

RIKER, WILLIAM H. (1962) *The Theory of Political Coalitions*. New Haven, CT: Yale University Press.

SEMMEL, ANDREW K. (1982) "Small Group Dynamics in Foreign Policymaking." In *Biopolitics, Political Psychology, and International Politics*, edited by Gerald Hopple. New York: St. Martin's Press.

STASSON, M. F., T. KAMEDA, C. D. PARKS, S. K. ZIMMERMAN, AND J. H. DAVID (1991) Effects of Assigned Group Consensus Requirements on Group Problem Solving and Group Members' Learning. *Social Psychology Quarterly* **54**:25–35.

STEIN, JANICE GROSS (1990) Real Time and Psychological Space: Decision Units, Decisions, and Behavior in Israel, 1967. Paper presented at the annual meeting of the International Studies Association, Washington, D.C., April.

STEIN, JANICE GROSS (1991) "Inadvertent War and Miscalculated Escalation: The Arab-Israeli War of 1967." In *Avoiding War: Problems of Crisis Management*, edited by Alexander L. George. Boulder, CO: Westview Press.

STEIN, JANICE GROSS, AND RAYMOND TANTER (1980) *Rational Decision Making: Israel's Security Choices, 1967*. Columbus: Ohio State University Press.

STERN, ERIC K. (1997) "Probing the Plausibility of Newgroup Syndrome." In *Beyond Groupthink: Political Group Dynamics and Foreign Policymaking*, edited by Paul 't Hart, Eric K. Stern, and Bengt Sundelius. Ann Arbor: University of Michigan Press.

STERN, ERIC K., AND BENGT SUNDELIUS (1992a) "The U-137 Incident: A Narrative Description." In *Krishantering: U-137-Krisen*, edited by Bror Stefenson. Stockholm: Royal Academy of War Sciences.

STERN, ERIC K., AND BENGT SUNDELIUS (1992b) Managing Asymmetrical Crisis: Sweden, the USSR, and the U-137. *International Studies Quarterly* **36**:213–239.

STERN, ERIC K., AND BERTJAN VERBEEK (1998) Whither the Study of Governmental Politics in Foreign Policymaking? A Symposium. *Mershon International Studies Review* **42**:205–255.

SUNDELIUS, BENGT (1989) *The Committed Neutral: Sweden's Foreign Policy*. Boulder, CO: Westview Press.

SUNDELIUS, BENGT (1990) Whiskey on the Rocks: Sweden's Response to a Trapped Soviet Submarine. Paper presented at the annual meeting of the International Studies Association, Washington, D.C., April.

THOMPSON, L. L., E. A. MANNIX, AND M. H. BAZERMAN (1988) Group Negotiations: Effects of Decision Rule, Agenda, and Aspiration. *Journal of Personality and Social Psychology* **54**:86–95.

VERTZBERGER, YAACOV Y. (1990) *The World in Their Minds*. Stanford, CA: Stanford University Press.

WALKER, STEPHEN G. (1990) British Decision Making and the Munich Crisis. Paper presented at the annual meeting of the International Studies Association, Washington, D.C., April.

WALKER, STEPHEN G. (1995) British Decision Making and the Munich Crisis, II. Mershon Center, Ohio State University.

WALKER, STEPHEN G., AND GEORGE L. WATSON (1989) Groupthink and Integrative Complexity in British Foreign Policymaking: The Munich Crisis. *Cooperation and Conflict* **24**:199–212.

# Foreign Policy by Coalition:

## Deadlock, Compromise, and Anarchy

### Joe D. Hagan
*West Virginia University*

### Philip P. Everts
*Institute for International Studies, Leiden University*

### Haruhiro Fukui
*University of California, Santa Barbara*

### John D. Stempel
*Patterson School, University of Kentucky*

When ultimate authority in foreign policymaking is neither a predominant leader nor a single group, there is a third alternative decision unit: a "coalition" of politically autonomous actors. The defining feature of this type of decision unit is the absence of any single group or actor with the political authority to commit the state in international affairs. Foreign policy decision making in these settings is very fragmented and centers on the willingness and ability of multiple, politically autonomous actors to achieve agreement to enact policy. One premise of this essay is that, although typically ignored in the study of foreign policy decision making, coalition decision units are actually quite prevalent across a variety of institutional settings. They are prone to occur in parliamentary democracies with multiparty cabinets, in presidential democracies with opposing legislative and executive branches, in authoritarian regimes in which power is dispersed across factions and/or institutions, and finally in decentralized settings in which bureaucratic actors gain authority in collectively dealing with major policy issues.

This essay's other premise is that coalition decision units—despite the fragmentation of political authority within them—are in fact able to produce a variety of decision outcomes. Drawing upon theories of coalition formation, we propose a variety of political variables that facilitate or inhibit the achievement of agreement in coalition decision units. Key among these variables is the nature of the decision rules that govern the interaction among coalition members in the policymaking process. Decision rules define three basic coalition configurations and are illustrated in some detail using case studies:

1. A multiparty coalition cabinet with an established decision rule that requires unanimous agreement as exemplified in the decision making of the Dutch government in the 1980s regarding the question of accepting NATO cruise missiles (Everts).
2. A largely interbureaucratic decision where the established decision rule requires only a majority vote as exemplified in Japanese decision making surrounding the 1971 "Nixon shocks" and, in particular, the pressure to devalue the yen (Fukui).
3. A revolutionary coalition in an authoritarian regime with no decision rules as exemplified in the case of Iranian decision making concerning the American hostage crisis starting in 1979 (Stempel).

These cases illustrate the interplay of variables that predispose coalition decision units to act in a variety of ways, ranging from the immobilism of extreme deadlock to the aggressiveness reflective of near political anarchy. As with the other pieces in this special issue, these cases provide an initial, detailed application of the theoretical logic linking coalition decision structures and processes to foreign policy.

## COALITION DECISION UNITS AND FOREIGN POLICY: A THEORETICAL OVERVIEW

Coalition decision units have two defining traits. One is the sharp fragmentation of political authority within the decision unit. No single actor or group has the authority to commit on its own the resources of the state; a sustained policy initiative can only be enacted with the support (or acquiescence) of all actors within the decision unit. Any actor in the decision unit is able to block the initiatives of the other actors. This may occur by (1) executing a veto, (2) threatening to terminate the ruling coalition, and/or (3) withholding the resources necessary for action or the approval needed for their use. Furthermore, for a set of multiple autonomous actors to be the authoritative decision unit, the decision cannot involve any superior group or individual that acts independently to resolve differences among the groups or that can reverse any decision the groups reach collectively.

The other defining feature of a coalition decision unit centers around the effects that each actor's constituencies can have on members of the decision unit. Even if representatives of the different actors within the coalition do meet (say, in a cabinet), these individuals do not have the authority to commit the decision unit without having first consulted the key members of those they represent. The power of these leaders is, in effect, incomplete since it can be significantly restricted by the views of constituents. Such constraint greatly complicates the ability of a coalition of actors to achieve agreement. For individual decision makers in this type of decision unit, the political process is, itself, a "two-level game" (Putnam, 1988) in which each decision maker must negotiate not only with opposing actors within the decision unit but also with factional leaders in his or her own constituency. As a result, foreign policymaking within coalition decision units reflects the bargaining that is ongoing within two domestic political arenas. Coalition decision units are, thus, constrained in what they can do.

The fragmentation of authority characteristic of coalition decision units is likely—but not automatic—in a wide variety of institutional settings (see Hagan, 1993). Indeed, such decision units can be found in all types of political systems. They occur in democratic and authoritarian regimes as well as in well-established and less institutionalized regimes. Consider the following four settings:

***Multiparty cabinets in parliamentary democracies.*** Coalition decision units may occur if no single party—or faction—has sole control of the cabinet due to the fact that none has an absolute majority in the parliament. At any time, the defection of a party (or faction) can bring down the cabinet which, in some cases, may even require new elections. As such, any party or faction may block the actions of the rest of the cabinet by threatening to defect from the coalition. In order for a foreign policy initiative to be taken, all members of the coalition must agree.[1]

---

[1] On the prevalence of coalition cabinets in parliamentary democracies in postwar Europe see Bogdanor (1983) and Lijphart (1984); each makes the point that two-party "majoritarian rule" hardly fits the norm in Europe or elsewhere (e.g., India, Israel, Uruguay, or, now, Japan). The literature on Japanese foreign policymaking within the factionalized Liberal Democratic Party is particularly rich (e.g., Hellmann, 1969; Destler et al., 1976; Hosoya, 1976; Ori, 1976; Fukui, 1977a). The work on other advanced democracies is a bit more scattered but some emphasizing political decision making include analyses of the Netherlands (Everts, 1885), the Scandinavian countries (Sundelius, 1982; Goldmann, Berglund, and Sjostedt, 1986), as well as Germany, France, and/or Britain (e.g., Andrews, 1962; Hanrieder, 1970; Morse, 1973; Hanrieder and Auton, 1980; Smith, Smith, and White, 1988). See also comparative theoretical studies by Risse-Kappen (1991), Hagan (1993), and Kaarbo (1996).

***Presidential (and semi-presidential) democracies in which the executive and legislature are controlled by opposing parties.*** Although separation of powers arrangements mean that presidents are not dependent on the legislature for retaining office, the executive normally shares significant policymaking authority with a similarly autonomous legislative branch. Because the two institutions have the ability to check each other's policy actions (without bringing down the government), a foreign policy initiative involving major commitment normally requires that the separate institutions must work together if substantively meaningful action is to be taken.[2]

***Authoritarian regimes with power dispersed across separate factions, groups, or institutions.*** Like parliamentary cabinets, one-party regimes, military juntas, and traditional monarchies may become fragmented with the presence of well-established and politically autonomous factions—each of which is essential to the maintenance of the regime's authority or legitimacy. More extreme fragmentation can occur in periods when such governments are in political flux (e.g., during periods involving revolutionary consolidation or institutional reform or, even, decay) and power is spread across the separate institutions typical of authoritarian regimes: the ruling party, government ministries, and military apparatus (Perlmutter, 1981). Whatever the case, foreign policymaking will reflect the interplay among these separate actors and the agreement (or lack thereof) among them.[3]

---

[2] On the role of the U.S. Congress as, in effect, a part of a coalition decision unit with the executive branch see Frank and Weisband (1979), Destler, Gelb, and Lake (1984), Destler (1986), and Lindsay (1994), as well as cases in Lepper (1971), Spanier and Nogee (1981), and Snyder (1991: ch. 7). LePrestre (1984) offers a useful analysis of French foreign policymaking during that country's first period of "cohabitation." Foreign policy decision making in Latin American states, many of whom have presidential regimes, is considered in Lincoln and Ferris (1984) and Munoz and Tulchin (1984).

[3] The literature on the politics of Soviet foreign policymaking was particularly rich, although arguably the Soviet Union did not decay into coalition decision making until the latter part of the Gorbachev regime. Works that highlight the dispersion of power in Soviet policymaking include those by Aspaturian (1966), Linden (1978), Valenta (1979), and Gelman (1984). Of course, Russian foreign policymaking now approaches that of being semi-presidential. The People's Republic of China is currently the key communist case; in the post-Deng era, there is good reason to believe that political authority has become quite dispersed in what is otherwise an established regime (e.g., Barnett, 1985). Cases of decaying authoritarian states are numerous. In addition to the historic cases considered in the first piece in this special issue (particularly Snyder, 1991), detailed studies of extreme decision-making conflict (or anarchy) and foreign policy can be found for revolutionary France (Walt, 1996), Sukarno's Indonesia (Weinstein, 1976), Syria prior to the 1967 War (Bar-Simon-Tov, 1983), China during the

***Decentralized interbureaucratic decision making.*** In all regime types with complex political organization, coalition decision units may emerge when the political leadership permits an issue to be handled in a decentralized setting. Power then gravitates to bureaucratic actors and even interest groups who interact on a more or less equal and autonomous basis. Cooperation among these actors is necessary because their decision must ultimately be sanctioned by the political leadership, and failure to resolve issues on their own risks outside political intervention.[4]

As the reader can see, because of the rather widespread fragmentation of institutional and political authority, coalition decision units are actually quite prevalent. That is not to say, however, that such fragmentation automatically leads to foreign policymaking by coalition. In settings in which there are norms that facilitate policy coordination among representatives so that they can work as a single group or when leaders are so deadlocked that a single individual or bureaucratic actor can achieve de facto control of an issue, we may not find coalitions as the authoritative decision unit. But there are enough instances where coalitions may be present to warrant examining their effect on the decision-making process.

## *Factors Affecting Agreement Among Autonomous Actors*

The basic theoretical task we have in linking coalition decision units to decision outcomes is to understand the process whereby separate and autonomous political actors can come together to take substantively meaningful actions in foreign affairs that are authoritative and cannot be reversed. The fragmentation of authority inside a coalition decision unit necessitates that a sequence of questions be asked in developing the explanatory logic for this type of unit. First, what kinds of resources count in shaping who had influence within the coalition, and how much of that resource is adequate to authorize a particular course of action? Second, what conditions lead separate, often contending, actors to achieve agreement on foreign policy? One's initial inclination would be to assume that such fragmented decision bodies find themselves internally deadlocked and unable to act. Although deadlock (in various forms) is an important outcome here, our assumption is governments with coalition decision units can

---

cultural revolution (Hinton, 1972), and Argentina and the Falklands (Levy and Vakili, 1990) as well as Iran and the hostage crisis (Stempel, 1981).

[4] The original literature on "interservice rivalries" within the U.S. military describes the classic case of this pattern of coalition decision making. See works by Schilling, Hammond, and Snyder (1962), Hammond (1963), Caraley (1966), Davis (1967), and Huntington (1968). Destler (1980) and Vernon, Spar, and Tobin (1991) illustrate a similar pattern with respect to foreign economic policy.

act in significant and meaningful ways. Indeed, as hypothesized with the other two kinds of decision units, the dynamics of the coalition decision unit may strongly amplify existing predispositions to act as well as diminish them.

As with the predominant leader and single-group decision units, we will draw here upon well-established theoretical research to conceptualize the dynamics regarding how coalition decision units can shape what governments do in the foreign policy arena. But, in marked contrast to the other two types of decision units, there does not exist a body of literature that explicitly and directly addresses the foreign policy decision making of politically autonomous actors. Even though some theoretical work has examined the foreign policy effects of organized opposition that is relatively proximate to the decision unit (e.g., Snyder and Diesing, 1977; Lamborn, 1991; Snyder, 1991; Hagan, 1993; Rosecrance and Stein, 1993; Peterson, 1996), it is necessary to turn to the field of comparative politics and, in particular, to "coalition theory" for a useful systematic body of empirically grounded theory. Although addressed to the larger question of government formation, the core theoretical concerns in this literature parallel ours. Like those involved in the development of theories of coalition formation, we seek to identify the conditions that facilitate agreement among autonomous and contentious political actors, none of whom has the resources needed to implement a political decision on their own, be it controlling a cabinet or authorizing a policy decision.

Throughout the coalition theory literature, there are two principal theoretical arguments about what motivates political parties to agree to join a multiparty cabinet. One of these is the "size principle" (Hinckley, 1981) which asserts that key to a player's behavior is its conservation of its own political resources. This principle is best embodied in the notion of the "minimum winning coalition" (Riker, 1962), which when applied to cabinet formation states that the number of parties in a coalition will total only enough to sustain a majority of seats in the parliament. Inclusion of additional parties would require a further distribution of resources (i.e., ministries) without any further gain to the parties already in the coalition. Similar logic applies to building support for a policy initiative within a coalition decision unit. Namely, agreement within a coalition decision unit will include only those supporters necessary for its acceptance by the entire body according to whatever voting or other decision rule may apply. Inclusion of additional actors is avoided because of the costs of (1) incorporating their preferences and thus making further compromises, (2) expending more resources in the form of side payments to uncommitted parties, and/or (3) sharing credit for a popular policy which may have the effect of enhancing the position of contenders for power in the regime. In making foreign policy choices, the conservation of political resources by each player rationally precludes including additional supporters in an agreement (e.g., compromise) if their support is not crucial to authorizing the state to a particular course of action.

The second principle in coalition theory is what De Swaan (1973) calls "policy distance." This principle underlies the "minimum range" theory found in the work of Leiserson (1966) and Axelrod (1970). The focus here is on the policy/ideological preferences of contending actors, with the assumption that rational "players wish to be members of winning coalitions with a minimal diversity" (De Swaan, 1973:75). Policy preferences are not intended to supplant Riker's concern for the weights and numbers of players. Rather, the two are combined as in Axelrod's (1970) conception of the "minimum connected winning coalition" in which a cabinet is expected to have a minimum number of parties who are also ideologically proximate. This elaboration on Riker's minimum winning coalition permits the proposition that agreements within coalition decision units will involve actors with relatively proximate preferences. For example, drawing upon Snyder and Diesing's (1977) and Vasquez's (1993) depictions of the broad policy divisions we often find in considerations of foreign policy, "accommodationalists" and "soft-liners" would be more likely to band together with each other than with distinctly "hard-line" elements.[5]

Although the principles of size and policy space form the core of coalition theory, the comparative politics literature has not stopped with these two concepts. Important empirical studies of cabinet formation have isolated major exceptions to the "minimum connected winning coalition" in postwar Western Europe, Israel, and Japan (see case studies in Browne and Dreijmanis, 1982; Luebbert, 1986; and Pridham, 1986). To account for these anomalies, additional factors have been suggested, including actors' willingness to bargain (Dodd, 1976), the presence/absence of a "pivotal actor" (De Swaan, 1973), the structure of party preferences (Luebbert, 1984), the level of information uncertainty (Dodd, 1976), the existence of consensus-making norms (Luebbert, 1984; Baylis, 1989), and, at the other extreme, the complete absence of institutionalized decision rules (Druckman and Green, 1986). This research provides key insights relevant to understanding the operation of decision units. They are incorporated into the coalition decision unit model in two ways: (1) as additional factors explaining the likelihood of agreement among coalition actors or (2) as contextual factors that define decision-making rules and thereby condition the interplay among members of the coalition and the precise effects of size, polarization, and the other variables.[6]

---

[5] This "connectedness" across actors' policy positions characterizes not only minimum winning coalitions but also the oversized and undersized ones that we discuss below.

[6] To the best of our knowledge, there does not appear to be consensus or synthesis concerning the relative importance of—or interrelationship among—the specific contingencies in the coalition theory literature.

Among those factors affecting the chances of agreement, one particularly important refinement of the principles of size and policy space is De Swaan's (1973) notion of the "pivotal actor." A coalition member is pivotal on an issue "when the absolute difference between the combined votes (weights) of members on his right and of members on his left is not greater than his own weight" (De Swaan, 1973:89). Any policy agreement must therefore include this actor, and because it can play off alternative partners its preferences will likely dominate an eventual agreement. When such an actor does not have strong preferences on the issue it can shape the decision by mediating conflicts between players on both sides of the issue in exchange for side payments on other issues including regime maintenance.[7] Either way, this concept of a pivotal actor refines our notions of a minimum connected winning coalition by identifying more precisely the players necessary for policy agreement as well as another source of political pressure for overcoming deadlock among otherwise polarized groups.

Another factor facilitating agreement is the willingness of one group to accept side payments and, more dramatically, political logrolling. The coalition formation literature notes that often small, issue-oriented parties may join (and support) a government in exchange for control of a single ministry or policy issue (see Browne and Frendreis, 1980; Hinckley, 1981). According to Luebbert (1984:241), this kind of bargaining arrangement is possible if groups within the coalition have "tangential" preferences, that is, ones "that address different issues and are sufficiently unrelated so that party leaders do not consider them to be incompatible." A modification of this aspect of coalition theory is directly applicable to coalition decision making because it suggests the possibility of breaking deadlock among politically antagonistic contenders. Advocates of a policy may be able to buy off a strong dissenter with concessions critical to them on another issue, something that is especially likely in the case of a smaller, single-issue party with critical votes (e.g., the religious parties in Israel with their domestic concerns). The implications of side payments can also be seen in a larger light using the theoretical argument developed by Snyder (1991). With regard to logrolling, he makes the point that opposing actors may, in effect, offer each other payments that concern foreign policy issues. The implication

---

[7] This is especially important when an individual leader has an institutionally pivotal position in the regime, yet is not committed to a particular issue or, more dramatically, fails to assert the authority of his or her institutional position. Such behavior can create a political vacuum and lead to a de facto coalition policy arrangement. For example, among pre-WWI governments relatively passive and ineffective leaders were critical to the emergence of hard-liners in the governments of Germany (William II) and Russia (Nicholas II). Interestingly, just the reverse occurred in France, where President Poincaré was able to impose relative coherence on the normally weak and fragmented government of the Third Republic.

of his definition of the concept goes far beyond permitting agreement—instead, both sides implement their policies to the maximum degree even though their actions may be contradictory and/or overextend the state internationally. Such an outcome is the opposite of deadlock—one of overcommitment rather than failure to act.

A further variable—or actually, set of variables—affecting the agreement among coalition actors is their willingness to bargain with each other. "Willingness to bargain" assesses the degree to which there are "serious a priori constraints on parties which make them hesitant to negotiate or strike bargains" (Dodd, 1976:41). Constraints on bargaining include extreme distrust between parties, immediate competition for control of the government, and opposition to agreements from factions within coalition parties (see Dodd, 1976; Lijphart, 1984; Luebbert, 1984; Pridham, 1986; and Steiner, 1974). Intense distrust or severe political competition between (and within) actors may lead members of the decision unit to define a policy problem as a "zero-sum" political issue. If political fortunes outweigh substantive policy merits, even actors with relatively similar policy positions are not going to be willing to bargain with each other. At the other extreme, the existence of strong norms of "consensus government" (Lijphart, 1984) and "amicable agreement" (Steiner, 1974) can greatly facilitate the coming together of parties with otherwise strong policy differences. Indeed, as illustrated by the Swedish and Israeli cases in the preceding article on single-group decision units, coalition cabinets can function as a single group if there are strong norms of political trust, strong party discipline, and habits of cooperation across ruling parties.[8]

## *Decision Rules Define the Context for Coalition Policymaking*

The other way of incorporating these additional variables into our exploration of the coalition decision unit is by combining several of them into what the framework calls a "key contingency variable"—in this instance, decision-making rules. The premise here is that decision rules define the context in which the properties of coalition size, policy space, pivotal actor, and willingness to bargain interact to produce outcomes ranging from agreement to deadlock. As with the predominant leader and single-group decision units, the idea of decision rules permits us to identify the "contingencies" that, in turn, point to alternative states in which coalitions operate.

The theoretical primacy given here to decision-making rules requires further explanation. Decision rules are the general procedures and norms that mem-

---

[8] For discussions of single-group decision making in coalition cabinets see chapters in volumes on the Netherlands edited by Everts (1985) and on Northern Europe edited by Sundelius (1982).

bers of the decision unit recognize as guiding interaction within that authoritative body. They are the "rules of the game [defining] the set of players, the set of permissible moves, the sequence of these moves, and the information available before each move is made" (Tsebelis, 1990:93). They range from formal constitutional procedures to more informal norms of behavior dictated by deeply rooted cultural practices, widely accepted lessons of past political crises, or the like. Whatever their origin, these rules shape political interactions within the coalition decision unit and thus define the context within which the process of achieving agreement occurs.

Decision rules help us understand the possibility of agreement among autonomous actors in at least two ways. They stipulate precisely what constitutes an authoritative consensus within the decision unit, that is, the number of votes required to win a debate and have the government adopt an initiative. Knowledge of this variable enables us to consider the range of votes required to achieve a "minimum winning coalition" among a subset of actors within the coalition. The other insight provided by decision rules stems from the degree to which these political procedures and norms are well established, or "institutionalized" (Huntington, 1968; see also Hagan, 1993; Mansfield and Snyder, 1995). Well-established decision rules (whatever the precise voting procedures) make clear the decision mechanisms by which separate actors are brought together, what kinds of resources matter in weighting influence in the coalition, and how these weights are to be combined in arriving at an agreement. They provide, in other words, "information certainty" concerning the relative weights of each player's resources and the likely prior moves of each in a bargaining setting (Dodd, 1976:40). Beyond this, and in more subtle ways, knowledge of the extent to which decision rules are established provides clues into the overall political relations among the actors within the coalition (e.g., a history of distrust, views on the nature of political relationships, and habits of cooperation). In these two ways even simple information about rules can tell us much about how the decision-making game is played.

The role given to decision rules here is not new. Their importance, as well as the institutions in which they are embedded, is found in several theoretical literatures. They are, of course, inherent in the coalition formation literature discussed above. Not only do they underlie Dodd's (1976) conception of "information uncertainty" as a constraint on coalition formation, but some of the starkest empirical exceptions to "minimum connected winning coalitions" have been found in highly consensual systems (e.g., Luebbert, 1984) and in very unstable polities (e.g., Druckman and Green, 1986). Decision rules are also prominent in the broader comparative politics literature, much of which accounts for anomalies in electoral and partisan behavior in different national political settings (e.g., Steiner, 1974; Lijphart, 1984; Thelen and Steinmo, 1992). Although widely associated with the notion of games in "multiple arenas" (as noted earlier), not to be forgotten is that part of Tsebelis's (1990) treatment of "nested games" that raises the complications in rational behavior that can stem from variations in institutional context and the fact that such differences may lead to changes in the rules themselves. Rational choice theorists generally acknowledge that the political context is critical to uncovering the logic underlying the strategies and preferences of political actors; in other words, that "individuals acting rationally can arrive at different outcomes in different institutional settings" (Lalman, Oppenheimer, and Swistak, 1993).[9]

In adapting ideas concerning decision rules to coalition decision units, we propose to differentiate among three general kinds of conditions under which coalitions may operate. Figure 1 diagrams the questions we seek to answer in deciding which kind of coalition decision unit we are observing at any point in time. The first question simply ascertains whether or not clear decision rules exist; it is followed by a second question that distinguishes between the voting requirements of non-unanimity and unanimity in those cases in which rules are well established. Although no claim is made here to have captured the many nuances extant in the literature about institutions, the interaction of these questions points to three general types of coalition decision settings. The middle path of Figure 1 conforms most directly to the dominant themes in coalition theory. This path describes the situation in which the decision unit is governed by established voting rules that permit an authoritative decision if a subset of actors (i.e., majority) achieves agreement on a particular course of action. It applies coalition theory's core notion of the "minimum winning connected coalition." But this path

---

[9] The centrality of decision rules is suggested in the international relations literature in "neoliberal institutionalism" (Keohane, 1984, 1989; Keohane, Nye, and Hoffmann, 1993; also Baldwin, 1993). Although a theory cast to explain international system dynamics (i.e., cooperation), the thrust of neoliberal institutionalism parallels our own and adds insights to those in comparative politics. As in the work of Keohane and other international relations theorists, our coalition decision-making model is concerned with explaining cooperation (or agreement) among autonomous players without any superior authority and in a condition of potential "anarchy." Keohane's (1984, 1989) analysis illustrates that across time egoistic actors not only can learn cooperation in conditions of stability, but also will develop certain "rules, norms, and conventions" that facilitate agreement. His functional argument is that a self-interested state will seek (or "demand") international institutions (or "regimes") for several reasons: they provide a clear legal framework establishing liability for actions, they provide information, and they reduce the costs of the transactions necessary for coordinated policies (Keohane, 1984). The same is likely true of actors within governments; well-established decision rules in a fragmented domestic political setting can facilitate agreement among contending actors for these same reasons. Kegley (1987) has, to our knowledge, attempted the most detailed and innovative application of "regime theory" to theorizing about the decision-making process. Our conception of decision rules is similar to his notion of "procedural" decision regimes.

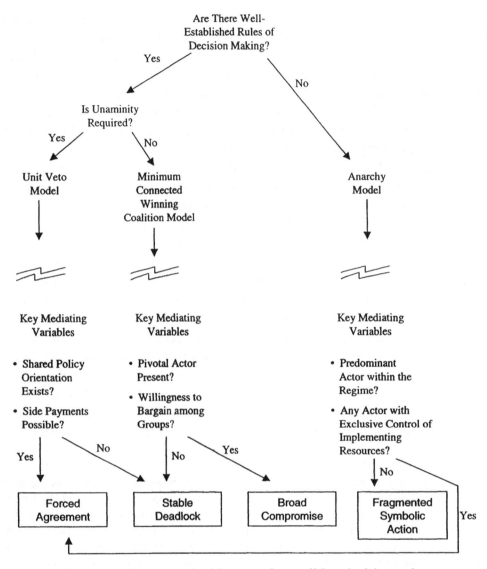

FIGURE 1. Summary decision tree for coalition decision units

does not encompass the full range of coalition decision units extant in different political systems, issue areas, and situations. For purposes of systematic, cross-national comparisons, it is imperative to consider two other politically extreme situations. One, which probably does not normally occur in the formation of governments (except during national crises), is where voting rules (or political necessity) require unanimity among all participants, in effect creating a "unit veto" system in which any single actor can block the initiatives of all others. This situation is diagramed on the left side of Figure 1. The other situation, found on the figure's right side, portrays the other extreme—essentially one of "political anarchy," in which established decision rules are largely absent and the overall political process is extremely fluid. There is maximum uncertainty about the political game, raising questions about not only the locus and allocation of authority but also the larger political stakes involved in the debate over policy.

## How Decision Rules Shape Policy

Having sketched the components and the logic behind the coalition decision unit, we now turn to applying the framework to specific episodes of foreign policy decision making by coalitions of autonomous actors. We will use the following cases to illustrate the three types of coalition decision units described above: the Netherlands and the 1979 NATO cruise missile crisis, Japan and the 1971 exchange rate crisis, and Iran and the 1979 U.S. hostage crisis. These cases are, respectively, examples of the unit veto model, the minimum winning connected coalition model, and the anarchy model as defined by the nature of their decision rules. The purpose of these cases is to explore the dynamics of coalition decision making in each type of configuration. Each case considers, how, first, decision rules condition the state of key coalition influences (again, coalition size, policy space, pivotal actors, and willingness to bargain) and, second, how, the rules shape the ways in which these factors interact to lead to a decision. The cases not only offer new insights into several non-U.S. cases but also lend initial support to our reinterpretation of coalition theory and what it indicates are important factors in determining how coalitions operate in different kinds of contexts.

### *Unit Veto Model: The Netherlands and NATO Cruise Missiles*

As envisioned in the unit veto model, coalition decision units have well-defined political rules that require agreement by all members to support any policy initiative or the decision unit cannot act. The imperative of "unanimity" stems from several factors. Constitutional arrangements may require that all parties formally commit to a particular course of action, for example, in the U.S. presidential system a declaration of war or implementation of a treaty requires the support of both the president and the Congress. Other imperatives might be less formal, yet equally compelling, in regimes in which executive authority is shared by multiple actors, as naturally occurs in parliamentary systems with multiparty cabinets. Although coalitions in stable parliamentary systems often function as single groups, sometimes issues are so politicized that they threaten to bring down the government as a result of party or factional defections from the cabinet (leading to a vote of nonconfidence). If an issue is so important that no actor is willing or politically able to allow itself to be overruled, then the government becomes incapable of action without bringing about its own collapse. Moreover, there are situations in which well-established norms require a consensus among participants if a decision is to be accepted as legitimate. These norms may be rooted in national political culture, as in Japan, but they may also be the result of institutional norms that have evolved among established actors, for example, those rules that govern interservice rivalries on U.S. defense budgetary decisions. Whatever the case may be, though, the common situation among these coalition decision units is that all actors must agree to support a decision if it is to take place. This situation is extremely fragmented because any actor alone can block the actions of all others, while any agreement must incorporate the full range of preferences within the decision unit. These conditions parallel those portrayed by Kaplan (1957) in his "unit veto" model of the international system.

One example of such a unit veto situation was the decision(s) by the Dutch government regarding NATO's deployment of a new generation of nuclear weapons in Western Europe in the early 1980s.[10] The issue was forced in December 1979 when the NATO ministerial meeting in Brussels, after long drawn out and difficult international negotiations, adopted the so-called double track decision. This agreement entailed the modernization of the alliance's nuclear forces with the deployment of 572 new intermediate range nuclear weapons, including 108 Pershing II missiles and 464 cruise missiles, from 1983 onwards. At the same time the NATO ministers called on the Soviet Union to begin arms control negotiations that could establish limitations on this category of nuclear weapons. The Netherlands was one of five European members of the alliance to receive the new missiles: 48 of the planned cruise missiles were to be deployed on its territory. The cruise missile issue would remain a severe policy problem for the succession of Dutch governments between 1979 and 1986.

In preparing its position with respect to the impending NATO decisions, the government of the Netherlands was subjected to sharp cross-pressures both from abroad and from inside the country.[11] From abroad, the Dutch government was under particularly strong pressure to agree to and cooperate with the intended program of nuclear modernization, not only from the United States but also from the Federal Republic of Germany (the latter did not want to become isolated by being the only country on the continent to have the new missiles on its soil and feared that continued Dutch opposition would boost the

---

[10] This section is condensed from Everts (1991) where the substance of the cruise missile issue and the application of the decision units model is presented in greater detail. The most extensive, although somewhat partisan, study of the decision-making process in the cruise missile case was written by one of the participating officials (van Eenennaam, 1988).

[11] Accounts of how consecutive Dutch governments fought hard and difficult battles to obtain concessions from their NATO partners are found in van Staden (1985), van Eenennaam (1988), and Soetendorp (1989).

morale of its own domestic opposition). Much of the leadership of the Dutch government were sympathetic to the arguments of these allies. The leadership shared the others' assessment of the need to counter the increased Soviet threat as well as the need for showing a united front to the Soviet Union in view of the outcome of the earlier debate on the neutron warheads. The ministers also worried about the possible effect of a negative deployment decision on the future credibility of the Netherlands within the NATO alliance. Refusal to accept the missiles would greatly undercut the position of the Netherlands within NATO, leading to accusations that this relatively small country was a "footnote country" or a "free rider" within the alliance. Should they back down on the deployment issue, ministers feared the Netherlands would risk losing their influence in future allied consultations about arms control.

Yet, domestically, the government's freedom to maneuver was constrained by a climate of antimissile public opinion that included not only parties on the Left but also relatively mainstream church groups opposed to (increased) reliance on nuclear weapons for security. Many shared the peace movement's opposition to all things nuclear (see Everts, 1983). Others, who did not reject all nuclear weapons, argued that to deploy new missiles in Europe would, if anything, lead to strategic de-coupling. Some argued that the new weapons would be not only superfluous but positively dangerous because they would increase the chances of nuclear war restricted to Europe and hence make such war more probable. It was also argued that deployment of the new missiles would fuel the arms race. The cruise missiles were seen as much more than a simple replacement of old weapons ("modernize your bicycle, buy a car," as one of the critics put it). If there was a Soviet threat (and most opponents shared this view), a further arms buildup was seen as a dangerous and ineffective way of dealing with it. The opposition was fanned by what was seen as dangerous loose talk by members of the Reagan administration on the possibility of a "limited nuclear war in Europe." These and other arguments of a more emotional nature would become the common stock of the mounting domestic opposition in the years following NATO's 1979 "double track" decision.

This polarized domestic political atmosphere greatly complicated Dutch foreign policymaking. Under normal conditions, foreign policymaking in the Netherlands was an elite affair in which critical issues were handled by a subgroup of cabinet ministers with responsibilities for foreign affairs.[12] However,

---

[12] This general procedure was not a small matter for countries that had multiparty systems and thus generally coalition governments. Like a number of European democracies (Baylis, 1989), well-established norms of interparty cooperation and intraparty discipline facilitated accommodation and permitted Dutch cabinets to operate as single-group decision units. See Lijphart (1968) for a more general treatment of norms of "accommodation" across the divisions in the Dutch political system.

this "classical" pattern of cabinet decision making was eroded to a considerable extent by the politicization of foreign policy issues during the period preceding the cruise missile episode. The democratization processes in Dutch society in general, which had taken place together with an erosion of the nation's postwar consensus on defense in the 1970s, had increased not only the desire of groups within society to participate in the foreign policy process, but also the wish of Parliament for a larger say in both decision making and control over the execution of foreign policy (see Everts, 1983). One result was that cabinets were now vulnerable to being removed from office as a result of votes of no-confidence supported by dissidents within ruling parties. In fact, in the years preceding the 1979 "double track" decision, there were no less than three times that a government crisis was threatened over an issue of foreign policy. That such events had happened hardly ever in the past is testimony to the new role of Parliament in the making of foreign policy and the relative weakness of the executive. On controversial issues such as the cruise missile deployment the cabinet could ignore the wishes of the Parliament only at its peril.[13]

The Dutch cabinets handling the cruise missile issue were internally fragmented and thus not in a good position to act decisively on this problem. On the one hand, the institutional positions of individual ministers (and therefore the party or faction each represented) are traditionally very strong. Dutch prime ministers are not predominant, technically serving merely as the chair of the Council of Ministers and in political practice as not much more than "*primus inter pares*." They cannot appoint ministers or force their resignation at will; the prime minister's role involves mainly the general coordination of government policies with respect to "politically sensitive" issues (even if he has strong views about the issues). And, on the other hand, the ability of the cabinet to act as a single group in making policy (and survive politically) is further complicated by the fact that Dutch governments are almost always multiparty coalitions. Because the electoral system is one of proportional representation none of the Dutch political parties has ever been able to win the majority of parliamentary seats necessary to rule alone, and therefore cabinets are formed by coalitions of two or more parties. Throughout the cruise missile crisis the Netherlands had a succession of coalition cabinets, that is, (1) from 1977 to 1981, a center-right coalition between the Christian Democrats (CDA) and a smaller Liberal Party (VVD), followed by (2) a center-left cabinet with the CDA, the Labour Party (PvdA), and Democrats '66 (left liberals) which lasted barely a year, and, finally, (3) a renewed center-right coalition with the CDA and VVD from 1982 to 1986.

---

[13] On the degree to which Parliament had become an independent actor in the foreign policy process at this time see the studies in Everts (1985).

Compounding the internal fragmentation of these cabinets was their wider vulnerability to being overthrown by Parliament on the cruise missile issue. Such was especially true in the case of the CDA-VVD center-right coalitions. These two cabinets, which were in power for all but one of the years of the cruise missile episode, had very small majorities in the Parliament. Indeed, the 1977–81 coalition had a majority of only two seats in the 150-member Dutch Parliament; the defection of just a few individual members could bring down the entire cabinet (as was threatened over foreign and defense issues several times in the late 1970s). Given these small majorities, it was highly probable that government positions on the question of the cruise missile deployment would play an important role in any parliamentary election. Public dissatisfaction with government policies could easily contribute to electoral defeats. And, in fact, the cruise missile issue played a major role in the 1981 general election campaign, the outcomes of which favored the left-leaning parties and resulted in the brief center-left coalition between the CDA and the Labour Party and Democrats '66. (Notably, although this center-left coalition had a larger parliamentary majority, it suffered from greater policy divisions and did not offer a durable alternative to the center-right CDA and Liberal Party coalition, despite its relatively large majority.)

Granting the severe constraints that fostered virtual deadlock among members of the Dutch cabinets on the cruise missile issue, are there any factors within coalition theory that, if present, might overcome these pressures for continued disagreement and promote some type of agreement within the Dutch government? Several present themselves. One is the presence of a pivotal actor in Dutch politics such as the Christian Democratic Party. This party was essential for many years to any cabinet, being both the largest of the Dutch parties and the only one able to transcend ideological issues.[14] But in this instance the CDA's ability to impose coherence on the cabinet was very limited because of its own internal fragmentation. In the period under consideration, the Christian Democrats were in the process of merging three parties (two Protestant, one Roman Catholic) into one new one—the CDA. At this time the merger was proving to be a difficult process. Marked divisions between the various groups

---

[14] The Dutch party system is characterized not only by a cleavage along the (socioeconomic) left-right continuum, but also by a cleavage between religious parties and more secular ones. While the latter is responsible for divisions on nonmaterial issues like education and abortion, the left-right dimension is otherwise dominant. Within this structure until recently the balance was maintained for all practical purposes by the Christian Democrats who held the center ground and as a result for a long time played a pivotal role in Dutch politics, being able to choose more or less freely with whom they would build a coalition. Center-right, and less frequently center-left coalitions (of various composition), always included the Christian Democrats.

along the left-right dimension were much in evidence, especially on foreign and defense policy. When the new cabinet was formed after the 1977 elections, eleven CDA members of Parliament (coming mainly from the former Protestant Anti-Revolutionary Party which would have preferred a center-left coalition) refused to give the new cabinet their formal blessing and reserved the right to judge it on its concrete policies. This group of "dissidents" selected a few issues on which they showed their "leftist" credentials. Probably because of their visibility and symbolic nature, these issues included a number of questions of foreign policy, such as the delivery of enriched uranium to Brazil, sanctions against South Africa, and neutron warheads. By using their implicit veto power and joining hands with the parties of the Left, which rejected modernization outright, the dissidents could make it very difficult—if not impossible—for the government to secure a parliamentary majority on the issue of cruise missiles. The "dissident" faction was, in effect, one of the actors within the Dutch coalition decision unit on this issue.

Another factor that might have promoted agreement among members of the Dutch cabinets is the possibility of what in coalition theory is called the formation of a diverse "national unity coalition" in the face of foreign peril or, in the case of domestic pressures, a common domestic strategy to ensure the government's survival. The cruise missile issue certainly posed severe pressures on the government of the Netherlands. As noted above, NATO and the Reagan administration were strongly pressuring the Dutch government to respond to the Soviet threat. And even if Dutch officials were not quite as alarmed as these other bodies by the Soviet threat, most were worried that the credibility of the Netherlands within the NATO alliance would be severely damaged. Yet domestic political pressures worked in exactly the opposite direction from what coalition theory predicts. The traditionally pro-NATO foreign policy elite and parties were under public and popular pressure *not* to accept the missiles. The leadership could not mobilize support behind the NATO commitment (e.g., the nationalism card was not viable), and approaching elections increased political pressures on the antinuclear opposition to maintain its credibility within the Parliament and the coalition. In short, domestic and international pressures were at odds with one another. Working as "cross-pressures," international and domestic pressures reinforced—not diminished—the divisions within the already divided coalition cabinet.

One final way suggested by coalition theory to achieve agreement within a coalition requiring unanimity is through side payments made by policy advocates to potential dissenters. Such a scenario would seem quite possible in the Dutch case given the key role of the small "dissident" faction within the Christian Democratic Party. However, even if these dissidents were willing to make concessions, they were not in a political position to do so because of their strong ties to the public opposition. These dissidents were under particular pressure from the churches, which could appeal to the fact that the new CDA party wanted explicitly to be a "Christian" party. They were an important asset for the church-sponsored peace organizations, IKV and Pax Christi, which played a central role in the societal debate over nuclear weapons in general. And they were central actors in the emerging broad coalition of groups, parties, and organizations opposing the deployment of new nuclear weapons. Having access to the churches, to which most CDA voters were attached, and being able to appeal to common religious ground, these peace organizations were able to make inroads into the political center. Agreeing to the deployment of cruise missiles at the height of the controversy would have destroyed the dissident faction.

In sum, the Dutch cabinets were severely constrained in dealing with the cruise missile crisis. Political elites were polarized over the cruise missile question and governing authority was fragmented to the extent that any action on the issue would bring down the government. Furthermore, factors that might have overcome these constraints all served to exacerbate—not dilute—them; for example, the pivotal CDA was divided, strong foreign and domestic pressures cut in opposite ways, and the dissidents were unwilling to accept side payments. The culmination of these pressures was largely a deadlock on the cruise missile issue. In subscribing to the communique of the 1979 NATO ministerial meeting, the Dutch accepted the arguments for modernization (but also stated that production of the missiles was an independent American decision), the idea of deploying 572 missiles in "selected countries," and their share of the common costs. At the same time they postponed their own decision on deployment within their borders. The basic reality was that the government could not act consistently in one direction or the other; it could not commit itself fully to either accepting the missiles or rejecting them. A single, coherent course of action was precluded. As a consequence, they engaged in minimalist foreign policy behavior regarding the cruise missile issue for a number of years.

All this discussion is not meant to suggest that there was complete disarray in Dutch foreign policy decision making. Although deadlocked, the tensions within the government were contained effectively in a way that reflected an orderly and sustained "papering over of differences," resulting in some diplomatic activity within the NATO context. In the ensuing parliamentary debate, cabinet ministers from both the CDA and VVD worked hard (and with some difficulty) to picture Dutch participation in the NATO agreement as substantively meaningful and yet satisfactory to all parties at home and abroad. The criticisms of CDA dissidents in Parliament were muted. Indeed, they finally recoiled at the prospect of a cabinet crisis, refusing to support a virtual motion of no confidence proposed by the opposition and merely restating that they did not accept responsibility for a deployment decision. The dissidents preferred to remain silent when the chips were down and the unity of their party was at stake. This behavior points to the further importance of the presence or absence of decision-making norms. It is striking that none of the parties in the Dutch government—despite their political differences—sought to appeal aggressively to the wider public or challenge the political system itself. This fact significantly enhanced the ability of the Netherlands to contain NATO alarm and pressures. The existence of strong consensus-making norms (Luebbert, 1984; Baylis, 1989) in Dutch politics enabled the deeply divided coalition governments to function with political restraint at home and abroad.

The deadlock over the deployment of cruise missiles finally came to an end in 1985. Further postponement of the decision was not possible beyond the NATO deadline of November 1, 1985. At that point the Dutch government decided to accept the missiles. Several factors contributed to this positive decision, even though the opposition had demonstrated one last show of strength with a petition drive requesting Parliament not to agree to deployment. This action was not sufficient to sway the attitudes of even the wavering CDA members of Parliament. The opposition had run its course and the left wing of the Christian Democrats had, for a variety of unrelated reasons, lost much of its strength within the decision unit. In other words, the pro-missile leadership of the Christian Democratic Party was now politically able to commit to the deployment of the missiles. The change in the decision unit was crucial to breaking the deadlock over the issue. However, because Dutch cabinets had well-established rules governing decision making, they had been able to handle their five-year deadlock in an orderly manner so that it remained contained and did not escalate into a wider political crisis either at home or within NATO.

## *Minimum Winning Connected Coalition Model: Japan and the Exchange Rate Crisis*

Not all coalition decision units with established decision rules require unanimity. Substantively meaningful foreign policy action can require agreement among only a subset of decision unit members. As noted earlier, it is this situation that is most directly analogous to the parliament-wide process of government formation in which only some parties enter the cabinet. In our case, established decision rules require that some kind of majority (one half, two thirds, etc.) reach a decision. In contrast to the setting requiring unanimity, it is important here to understand the relatively proximate preferences of the subset of actors that are in agreement and the processes that brought them together. Indeed, the outcome of this process presents the very real possibility of one side winning for the most part or, at the least, a compromise among actors with relatively proximate preferences. Whichever is the case, though, the decision will be defined by the need of participants to conserve influence over policy by making minimum compromises. In this way the logic of the minimum winning connected coalition underlies the dynamics of coalition decision units with established rules that do not require unanimity. This overarching argument can be supplemented by additional concepts suggested by coalition theory that further define, if not restrict, the evolution of agreement.

The case of Japan's 1971 exchange rate crisis illustrates the dynamics of coalition decision units with rules that are well established but do not require unanimity.[15] As in the Dutch case, this crisis brought severe international pressure to bear on the Japanese government. It was provoked by the August 15th speech by U.S. president Richard Nixon in which he announced a new and drastic domestic economic policy initiative, soon dubbed the New Economic Policy, which included the suspension of the convertibility of the dollar and the imposition of an across-the-board 10 percent import surcharge. The announcement was unambiguously intended to force the United States' major trading partners, especially West Germany and Japan, to help balance the U.S. trade and current accounts and to rescue its domestic economy from deepening recession by, above all, revaluing their own undervalued currencies. As immediately understood by most governments around the world, the announcement also sounded the death knell for the Bretton Woods monetary system which had supported the postwar international economic system for a quarter century.

This component of the "Nixon shocks" struck at the post-WWII mindset shared by the Japanese leadership—that is, the ruling Liberal Democratic Party, the powerful economic ministries, and the business community—concerning the existing international economic, especially monetary, system. Throughout the postwar period the Bretton Woods system of international monetary management and coordination had been highly beneficial to Japan as well as to West European economies and had helped them recover from the devastating effects of World War II within a remarkably short time and then achieve sustained growth at an unprecedented rate. The yen-dollar exchange rate set in 1949 at 360 yen per dollar (taken as part of the sweeping postwar economic reforms under American direction) increasingly undervalued the yen after the mid-1950s and thereby significantly added to the price competitiveness of Japanese exports in international markets. The 360 per dollar exchange rate thus served as a key stimulus and incentive for the expansion of Japan's export trade and the growth of its export-oriented industries (Shinohara, 1961, 1973:18–34; Yoshitomi, 1977:20–25). The mindset nurtured by this highly favorable experience predisposed the Japanese, both inside and outside government, to resist any suggestion of a significant change in the system, especially a revaluation of the yen.

---

[15] This case is explained in greater detail in Fukui (1989); see also Fukui (1979, 1987). The research is based on that author's interviews of key Japanese policymakers as well as accounts in Japanese newspapers and other publications. Due to space limitations these sources have been trimmed here but are available in the longer piece.

When the chronic and worsening trade imbalances with Japan and the European Economic Community nations led Washington to make such a suggestion in the late 1960s and early 1970s, Tokyo resisted and fought it. On the eve of the exchange rate crisis, the Japanese government was firmly set against a revaluation of the yen and, for that reason, also against a fundamental change in the Bretton Woods monetary system, both out of habit and for specific policy reasons. A departure from the Y 360 per dollar exchange rate would have seriously threatened Japan's key industries and devastated many of its export-dependent small businesses. Inasmuch as the key industries were the largest and most reliable contributors to the political funds of the ruling Liberal Democratic Party (LDP) and small businessmen were among its staunchest supporters at the polls, the anticipated impacts of a yen revaluation on their fortunes was a cause for serious concern in the cabinet and among some of the Ministry of Finance (MOF) and Bank of Japan (BOJ) bureaucrats (Sasaki, 1973:34).

Moreover, the mindset and predispositions created by twenty years of highly profitable experience under the Bretton Woods system was significantly reinforced in 1970 by a downturn in the domestic economy and the Japanese government's determination to nip the incipient recession in the bud. The government in Tokyo responded to the downturn with a series of reductions in the official discount rate beginning in October 1970 and a cabinet decision in March 1971 to speed up public works spending budgeted for the 1971 fiscal year (Okurasho, 1982:4). As Washington's call for a yen revaluation became more and more audible and insistent, Tokyo announced a program of eight specific defensive measures in early June. Known officially as the Program for the Promotion of Comprehensive Foreign Economic Policy, this eight-point program was unmistakably aimed at warding off the growing pressure for a revaluation of the yen by controlling Japanese exports and increasing imports, thus helping Washington balance its own trade account and hopefully get off Japan's back (Watanabe, 1981:198).

The exchange rate crisis was of paramount significance to the Japanese government. As such, it would be expected that the crisis would be handled by the country's senior political leadership, that is, the cabinet controlled by the leadership of the Liberal Democratic Party. And, in some respects, the LDP leadership was in a good political position to handle the crisis. In contrast to the Dutch cabinets above, the LDP was quite secure in its control of the cabinet. The LDP had been in power throughout the postwar period and had a sizable majority within the national legislature, the Diet (and at no time had dissident factions in Parliament engineered the downfall of the cabinet by a vote of no confidence). Yet the LDP was itself significantly fragmented by its own structural factions which were at this time like mini-parties that competed for control of the party's top leadership posts. Although insulated from parliamentary overthrow, competition among factions was intense and it was not unusual for cabinets to change as a result of policy failures or a shift in the balances among the factions composing the cabinet. As a result, Japanese foreign policymaking has typically involved great caution and a carefully crafted consensus on controversial issues—precisely the sort of constraints we might expect in a coalition of multiple autonomous groups.[16]

Japanese foreign policymaking in the exchange rate crisis did *not* involve the full LDP cabinet, even though that same cabinet dominated decision making on two other simultaneous issues—Okinawa reversion (Fukui, 1975; Destler et al., 1976) and recognition of the People's Republic of China (Ogata, 1988). Yes, the Japanese cabinet met immediately after the Nixon announcements, deciding that Japan should resist the U.S. move through multilateral consultations while implementing the eight-point stimulus expeditiously. Furthermore, a subsequent emergency meeting of the cabinet subcommittee, the Council of Economic Ministers, produced consensus decisions that sought to maintain the current exchange rate along with engaging in diplomatic consultations and initiating a domestic economic stimulus. But the actual locus of decision making moved to the Ministry of Finance and its various officials, including both of its senior ministers, their advisers, and the semiautonomous Bank of Japan. Despite the paramount political importance of this crisis, the prime minister and other LDP leaders adhered to a well-established rule of decision making within the Japanese government on issues that do not require legislative action; such issues are normally left to bureaucrats to decide, especially if the problems are highly technical. This tendency was reinforced by the current prime minister, Sato Eisaku, who not only was preoccupied with other issues with the United States (Okinawa and the textile dispute), but was an extremely cautious politician who would wait for consensus to form among his subordinates before he would act on any controversial policy issue (Kusuda, 1983).

Still, decision making within the Ministry of Finance did not take the form of a single group under the authority of Minister of Finance Mizuta Mikio. The de facto dispersion of power across separate actors (particularly the semiautonomous Bank of Japan) was apparent from the beginning of the MOF deliberations. In its initial meetings, soon after the August 16th cabinet meeting,

---

[16] In fact, most literature on Japanese foreign policy decision making highlights the severe constraints posed by factional politics within the Liberal Democratic Party cabinet. Along with the overviews cited in footnote 1, see case studies of security issues by Hellmann (1969), Fukui (1970, 1977b), and Welfield (1976) as well as of economic issues by Destler, Fukui, and Sato (1979) and Fukui (1979). This literature, particularly that dealing with economic issues, challenges the view that Japan's bureaucratic-dominated government acts, as a rule, with unity of purpose, consistency, foresight, and in the nation's best interest (Vogel, 1979; Johnson, 1982; Pempel, 1982).

senior MOF officials and a few BOJ representatives met to discuss and decide whether or not they should close the Tokyo foreign exchange market the next day. The officials were immediately divided over the issue. While the two top MOF officials, Vice Minister Hatoyama Iichiro and Deputy Vice Minister Hosomi Takashi, favored the closure of the market in order to avoid further confusion in the marketplace, the ministry's International Finance Bureau (IFB) officials opposed such action on the grounds that the market could not possibly be kept closed for more than a week without causing serious problems for trade transactions and that, when the market was reopened, the government would be put in a position where it would have to permit either a revaluation of the yen or the introduction of a floating rate system or both (Yamamura, 1984:140–141).[17] From the IFB officials' point of view, the only sensible thing to do under the circumstances was to keep the market open, implement the eight-point program, and wait for a multilateral solution to be worked out among the major industrial nations. MOF adviser Kashiwagi Yusuke and BOJ deputy governor Inoue Shiro sided with the IFB officials.[18] Since neither side was willing to change its mind, they agreed to present both views to Finance Minister Mizuta and let him decide. Mizuta first postponed the decision until the next morning, then decided to keep the market open, in effect supporting the IFB against the vice minister.

The initial decisions taken on August 16th point to several features of coalition decision making within the Ministry of Finance. First, note that the MOF coalition decision unit did take action and that these decisions did not require unanimity. Minister Mizuta agreed to actions resisting American pressure for devaluation in a way that overruled two of the top MOF officials—a vice minister and a deputy vice minister. In essence, the opinion of the "currency experts" in the ministry prevailed over the top amateur officials, as journalists observed a few years later.[19]

Second, Minister Mizuta's decision should not be taken to suggest that he was operating as a predominant leader. As minister of finance, he acted as arbiter between the contending groups of MOF and BOJ bureaucrats and ruled

---

[17] Also Yomiuri shimbun, August 18, 1971.

[18] The experts insisted on keeping the market open partly because of the overwhelming weight (92%) of the dollar in Japanese foreign transactions (Watanabe, 1981:198–199) and partly because Japanese banks authorized to handle foreign exchange had accumulated huge dollar reserves, estimated at the time to be worth about Y 1 trillion at the par value, under an export promotion program with the BOJ's deliberate encouragement. In light of the uneven distribution of the dollar holdings among the banks and for fear of giving the impression of unfairness in its treatment of them all, the BOJ also opposed the shift to a floating rate system (Yamamura, 1984:142–143).

[19] Asahi shimbun keizaibu, 1974:240.

in favor of the IFB experts' recommendation to keep the market open. To be a predominant leader he would have had to be able to make a decision on his own and then impose it on the entire government, including the MOF and BOJ bureaucrats. Mizuta was clearly not in such a position. He could not make a decision unless he was asked to do so by the bureaucrats and, even when he was asked to do so, was restricted to the options recommended by the bureaucrats. To have acted on his own would have violated the well-established rule of decision making in the Japanese government on issues not requiring legislative action mentioned above. In short, Mizuta acted, as any Japanese minister in a similar position would have, as an arbiter to choose between the two courses of action recommended by the bureaucrats, not as a predominant leader with freedom to choose any course of action he personally preferred.[20]

Third, decision-making authority ultimately was centered around autonomous bureaucratic elements within the Ministry of Finance. The well-established rules of the game required the involvement of experts in both the MOF and the BOJ as equals in practice, if not in theory. On strictly legal grounds, the cabinet or a subcommittee of it could have acted as a dominant group. If neither did then the MOF bureaucrats could have made most of the key decisions without the concurrence of their BOJ counterparts. Given the force of tradition and custom, however, and the strong consensual norm that pervades Japanese society, it would have been quite extraordinary for either the cabinet or the MOF bureaucrats to claim such a role (Richardson and Flanagan, 1984:333–336; cf. Krauss, Rohlen, and Steinhoff, 1984). As it turned out, neither did, and the MOF and BOJ bureaucrats were jointly responsible for all the key decisions, while a cabinet subcommittee, the Council of Economic Ministers, was nominally involved in the decision-making process. As will be seen below, the power of the Bank of Japan is particularly striking—even in comparison to other experts in the MOF. Although the BOJ is legally subordinate to the MOF, the two groups cooperate closely and make decisions by consensus. This procedure is partly due to tradition and custom and partly to the fact that the BOJ's contingent of experts, concentrated in its Research and Statistics Department, is both larger and, according to several insiders, more capable than its counterpart in the MOF's Research and Planning Division.

Japanese policy and decision making did not remain stagnant, even though the full cabinet and the Council of Economic Advisors confirmed these deci-

---

[20] The same applied to Mizuta's and BOJ governor Sasaki's actions on the shift to a floating rate system and the central issue of a yen revaluation. On both of these issues Mizuta and Sasaki simply ratified the consensus decisions previously reached among the bureaucrats.

sions the next day.[21] Within a couple of days (by August 19th), some business leaders and academic economists close to the government had begun to call for a revaluation of the yen, not because they preferred a floating rate system but because they believed the exchange rate system to be on the verge of collapse unless the yen was revalued (Takeuchi, 1988:205). While political leadership in the cabinet and the Council of Economic Advisors refused to yield to such a view and stuck to their commitment to defend the cheap yen, there was important movement elsewhere within the government among bureaucratic officials. As was the case earlier, key decisions were made by the MOF and BOJ officials and no politicians, except members of the Council of Economic Advisors, were involved in the process to any significant extent. In fact, it was from within the MOF-BOJ coalition that Japanese policy began to change.

The change in government policy began on August 22nd, when senior MOF officials, including Minister Mizuta, met in a secret conference where Director Sagami Takehiro of the Research and Planning Division of the Minister's Secretariat presented a policy paper that outlined in considerable detail what steps the Japanese government could and should take in coping with the deepening crisis. He began with the argument that, should Japan continue to stick to the eight-point program alone and persist in its insistence on the maintenance of the existing dollar-yen exchange rate, it would only be isolated from the international community; therefore, Japan had really no choice but to change its policy (Yamamura, 1984:149). He then suggested that, theoretically, four policy alternatives were available: (1) increase the range of exchange rate fluctuation permitted under the existing fixed rate system (1% in either direction), (2) adopt a dual rate-system (as France had done), (3) unilaterally revalue the yen by either 10 percent or 15 percent, or (4) adopt a floating rate system. He proceeded to point out that neither alternative (1) nor (2) would help solve the problem at all, but alternative (3) with a 10 percent increase in the value of the yen was worth considering while a 15 percent increase would have too deflationary an impact on the economy, and, finally, alternative (4) might also

---

[21] On the afternoon of the 17th, the regular cabinet meeting was followed by an emergency meeting of the cabinet subcommittee, the Council of Economic Ministers, where consensus decisions were made on several pressing issues: first, that exchange rate adjustments should be sought through a multilateral consultation at the forthcoming meeting of the Group of Ten (G-10) finance ministers; second, that the yen's current par value (i.e., $1 = Y 360) must be maintained at all costs; and third, that the eight-point program must be fully implemented (Okamoto, 1972:324). The ministers also decided to send a government representative to Europe and the United States on an information-gathering mission. MOF adviser Kashiwagi was appointed as the emissary and left Tokyo the following day, the 18th, for Paris, from there to proceed to London and Washington. Before he left Tokyo he told his MOF colleagues not to close the market until he returned, probably on the 23rd.

be worth considering. This secret MOF meeting apparently began with the consensus that a yen revaluation was unavoidable and ended with a new consensus that the adoption of a floating rate system was also unavoidable.[22]

The new MOF consensus was conveyed to the BOJ leadership the next day, Monday, the 23rd. The BOJ's senior officials, mainly from its Foreign and Business Departments, met and reaffirmed their support of the fixed rate system ostensibly on grounds that a floating rate system would seriously interfere with transactions between Japanese exporters and foreign importers (Yamamura, 1984:150–151). They agreed, however, that it would not be possible to maintain the fixed rate system indefinitely and that, moreover, the MOF had the right to change the yen's par value in any case with the BOJ having no choice but to accept the MOF's decision. Then, early on the morning of the 24th, BOJ governor Sasaki Tadashi and MOF minister Mizuta met. Sasaki expressed his personal support for the shift to a floating rate system but asked for a few additional days, until the 28th, before a final decision was announced, so that he might bring opponents among BOJ officials around. The two agreed to a five-point memorandum of understanding: (1) adoption of a floating exchange rate system in the spirit of international cooperation; (2) implementation of the new system at the earliest possible date, but no later than the end of the week, that is, Saturday, the 28th; (3) issuance of a government statement upon the implementation of the new system; (4) intervention by the BOJ in the foreign exchange market in order to prevent speculative transactions; and (5) BOJ intervention when the yen rose by a certain percentage (about 7–8%).[23]

This interministerial agreement reflected the recognition by MOF and BOJ officials of the futility of further resistance to accepting a floating rate and, as such, reflected the preferences of key MOF and BOJ officials—ones that overrode opposition by a minority of bureaucrats as well as the Council of Economic Ministers. Yet it was a broad compromise. Not only was the decision not implemented until August 28th, but the joint announcement by Mizuta and Sasaki stemmed from mutual concessions: it referred to the immediate shift to

---

[22] MOF adviser Kashiwagi, one of the original decision makers, returned to Tokyo on the 23rd and reported that same night to the meeting of senior MOF officials, including Mizuta, that most European governments would refrain from acting hastily on their own and would wait for a multilateral forum to agree on a common action. He also reiterated his view that Japan should not unilaterally revalue the yen (Asahi shimbun, August 24, 1971; Uchino, 1976:514). Kashiwagi's view, however, was clearly a minority opinion now and failed to have much effect on the new consensus that had emerged among the ministry officials during his absence.

[23] The space for the percentage was left blank, but Director-General Takeuchi Michio of the MOF Minister's Secretariat, who attended the meeting, jotted down the numbers in his minutes of the meeting.

a floating system in points 1 and 2 in deference to the MOF's new position, while it also referred in points 4 and 5 to the BOJ's intention to continue to intervene in the foreign exchange market when the expected change in the yen-dollar exchange rate reached a certain magnitude. Point 3 about the timing of the official announcement of the MOF/BOJ compromise decisions was obviously decided by consensus. This compromise was then ratified by the cabinet ex post facto. In their simultaneous but separate press conferences, Mizuta and Sasaki both emphasized the provisional character of the decision and their intention to maintain tight controls over the foreign exchange market and to let the BOJ continue to intervene whenever necessary, that is, not only to practice a dirty float but also to try to return eventually to a fixed rate system.

In sum, although the "dirty float" was extremely expensive for the Japanese government, this compromise avoided deadlock and enabled the fragmented Japanese government to adjust relatively quickly to new realities in the international economy by abandoning its prior consensus supporting fixed exchange rates. Several factors permitted this compromise to occur. One was that established decision rules did not require unanimity within the coalition decision unit. Thus opponents within the unit could be overridden (e.g., the BOJ) or were, in fact, overruled (e.g, the minority of bureaucrats led by Kashiwagi). Another was the range of positions on the issue, that is, those officials who recognized the need to adjust had relatively proximate positions and were willing to bargain with one another. Finally, the nation's political leadership in the cabinet was willing to defer the decisions to the ministerial experts, and as a result the decision did not invoke the factional deadlock that often occurred on other issues. These well-established norms not only narrowed the range of policy positions but also served to depoliticize the issue so that parties were willing and able to bargain without appealing to supporters in the wider political arena. All this is in sharp contrast to the deadlocked Dutch government where U.S. missiles threatened the government's survival. However, like its Dutch counterpart, the established decision rules in the Japanese government served to contain the crisis. Such cannot be said of our third case: Iran's handling of the American hostage crisis.

## Anarchy Model: The Iranian Hostage Crisis with the United States

The defining feature of this case is the near total absence of accepted, basic rules for decision making, a situation that is typical of less-institutionalized political systems (see Hagan, 1993: ch. 2). The absence of accepted decision rules greatly complicates the policymaking process, especially when power is dispersed across actors within a coalition decision unit. This complexity is manifest in several basic ways, each of which extends Dodd's (1976) notions regarding information uncertainty to far greater extremes than found in any Western democracy. First, as suggested by Druckman and Green (1986) in their coalition analysis of post-Marcos regimes in the Philippines, the absence of established procedures such as voting creates fundamental uncertainties about what kinds of political resources (force, legitimacy, economic benefits) are necessary to influence the nature of any decision and in what amounts. Second, this uncertainty suggests that when an agreement does emerge it will reflect a much less precise process than is found in conventional cabinet coalitions. In particular, without voting rules it is unclear how resources are combined to reach a decision. The result could easily be an oversized coalition as advocates of a policy rationally ensure against the political uncertainties on which the decision rests (Dodd, 1976). Third, as Tsebelis (1990) contends, decision conflicts are likely to become fights over the shaping of the rules themselves. Where institutions are weak or absent, the nesting of policy and political strategies can be severe, sharply intensifying and transforming the nature of the political game. The potential for each coalition actor to use foreign policy as a means of political survival (e.g., aggressively resorting to nationalism) is a viable option in such situations.

Iran's handling of the "hostage crisis" with the United States offers a clear example of decision making by a coalition decision unit that has few, if any, established rules or procedures for making decisions.[24] Not unlike other revolutionary regimes (see Walt, 1996), the seizure of the American embassy in November 1979 posed a severe threat to Iran's new revolutionary regime right at the time when alternative constitutional arrangements were first being proposed. The seizure of the embassy by radical student militants operating on their own was not, of course, an action of the government. Although viewed in different ways by moderate and hard-line elements within the revolutionary coalition, it was clear to all that the implications of the hostage situation were considerable. A crisis with the United States jeopardized trading relations with the West and exacerbated difficulties already apparent in the declining postrevolutionary economy. Furthermore, it increased the revolutionary regime's international isolation and even created the possibility of an American intervention. Equally severe were the domestic political implications of the crisis. Not only did the students' actions defy government authority, the symbolism of holding the embassy and confronting the Carter administration went straight to the heart of the question of legitimacy raging at the time in a highly volatile Iranian political system. Conciliation on this issue could, quite simply, undercut the

---

[24] Analysis of this case draws directly on Stempel (1989) and, in turn, upon Stempel's (1981) book-length study, *Inside the Iranian Revolution*, as well as on the various sources cited throughout the description of the case.

revolutionary credentials of the new regime as well as the political position within the government of any proponents of such moderation; that is, those who wanted to release the hostages were "not truly Islamic" (the clerical right) or "toadying to the Western imperialists" (Marxist left).[25] As such, the crisis was intertwined with the vulnerability of the regime, the fragmentation of authority within it, and the evolution of the revolutionary regime—including the precise institutions and norms in its still evolving constitution.

The absence of any constitutional order sharply magnified all of these pressures. At the time of the hostage seizure the situation within the Iranian government—then, the Provisional Revolutionary Government (PRG) headed by moderate nationalist Mehdi Bazargan—was one of extreme flux; that is, not only was a constitution yet to be adopted, but the range of legitimate political activity was being progressively narrowed. Originally, the PRG, as the Ayatollah Khomeini's designated government, had had widespread support among almost all of the public and virtually all organized political groups, which had given it the authority to implement revolutionary changes, dismantle the Shah's political order, and fundamentally restructure the historically close relationship with the United States. However, after just a few months in power, the political position of the PRG began to deteriorate at all levels with sharp divisions emerging over various policy issues, including the increasingly politicized issue of relations with the West. In particular, the PRG faced opposition from radical Islamic clerics among the broader revolutionary leadership (but not members of the government itself). Adopting an exclusionary political strategy, these radical clerics began to pick away at both moderate and leftist forces at all levels of the political system, including those in the original revolutionary coalition. First to leave (in April 1979) the government were the National Democrats and then the National Front, both Western-focused organizations seeking to reshape Iran's government in the European social-democratic model.[26] Similarly, leftist elements of the coalition such as the People's Fedayeen and other radical Marxist groups were driven underground because of their opposition to clerical domination in the shaping of "Islamic socialism" and various other cultural and constitutional issues.[27] Not long after the outbreak of the crisis, when it became apparent that Khomeini would not overrule the students, the

---

[25] A brief description of this interplay can be found in Richard Cottam's opening chapter in Ramazani (1990).

[26] The National Front was the party of former Prime Minister Mossadeq. Its leader, Karim Sanjabi, was the foreign minister in the moderate Bazargan's first cabinet while the leader of the National Democrats was Mossadeq's grandson.

[27] By contrast, the more Islamic (as opposed to Marxist) Fedayeen and Mujahidin groups continued to play important roles, sliding in and out of support for the Provisional Revolutionary Government.

moderate Bazargan resigned and Khomeini transferred authority to the newly created Revolutionary Council (Stempel, 1981:226).

Chaos and fragmentation were the hallmark of decision making *within* the struggling revolutionary coalition consisting of the Ayatollah Khomeini, moderates in control of government ministries, and hard-line clerics lodged in the new Revolutionary Council. Despite his prominence as the unchallenged leader of the Islamic revolution, Khomeini did not act as a predominant leader. His policymaking role was one of disinterested aloofness. As the *Velayat-eh-Fagih*, or Supreme Jurisprudent, who acts as the guardian of state authority according to his own theory of Islamic government, Khomeini deliberately kept himself insulated from day-to-day politics. Although he would intervene on occasion to ratify policy decisions (or "nondecisions"), he was not a decision maker in the sense of a single predominant leader or even an active participant in the policymaking process. His role might best be described as that of a court of last resort. When political conflict became too intense or threatening to the regime (and its legitimacy), Khomeini would decide on a politically acceptable policy line which all then followed—at least until they tried to reverse it the next time. He was, at most, a passive—but never entirely absent—member of a larger decision-making coalition.

No single actor—neither individual nor group—within the revolutionary coalition was capable of filling the void created by Khomeini's political style. On the one hand, no institutional entity existed with the clear authority to deal with the hostage crisis. In fact, as we have already observed, soon after the hostages were seized, Khomeini replaced the PRG with a new structure—the Revolutionary Council—which was to be the supreme authority in the Iranian regime. Its own power was, however, never fully established. Not only did Khomeini retain influence, but the Revolutionary Council had to share authority with other government entities particularly after a new constitution created a separate office of the president. Moreover, the Revolutionary Council was broadly split into two increasingly polarized political groups: the radical clerics of the Islamic Revolutionary Party and the relatively moderate government officials left over from the Provisional Revolutionary Council. Neither group was willing or able to dominate the Revolutionary Council. Furthermore, decision-making rules such as voting procedures were not well established; indeed, increasingly the very membership of the Revolutionary Council was open to question.

The largest group represented in the Revolutionary Council was the Islamic Republic Party (IRP), the most consistent and continuing player during the entire crisis. Though composed of many factions, it included most of the revolutionary coalition's religious elements and was led by radical clergy such as Ayatollah Behesti, Ayatollah Montazari, Hojotallah Ali Akbar Hashemi Rafsanjani, and Ayatollah Khomeini. The IRP's strength came from its senior clerics who, through their own feudal, quasi-bureaucratic networks, gradually acquired control of certain government ministries as well as most nonstate organizations. They had the support of Khomeini. Though they could not agree on questions regarding private property, the export of the revolution, and the relative evil of the U.S. and USSR, their common interest in holding power was strong enough to give the party political clout. They did, however, have internal consensus concerning commitment to Islamic government, rejection of Iranian nationalism in favor of pan-Islamic goals, and an intention to adhere to a "no East, no West" foreign policy which seeks economic self-sufficiency and supports Third World liberation movements and terrorism as state policy.[28] Furthermore, within the IRP there was a shared distrust of other groups in the revolutionary coalition; as a result, it became the driving force behind the regime's exclusionary political strategy.

Despite its power, the IRP was not able to dominate the revolutionary coalition and government institutions, but had to share with moderates and secular religious nationalists. These individuals were originally grouped around Prime Minister Bazargan's Liberation Movement of Iran and the coalition leadership of the PRG. They wanted an Islamic Republic that would uphold democratic values, in contrast to the radicals' authoritarian ideology. They also favored a foreign policy that put Iran's national interest first, and sought some accommodation with the West. Bani Sadr, who was elected president of Iran in January 1980, also fell into the secular wing of the revolutionary coalition, although he was not a moderate like Bazargan but a leftist Islamic academic figure. Though he originally favored releasing the hostages, he did so because he wanted to establish an Islamic Marxist state. The power of these more moderate and secular elements stemmed from several sources. They had well-established reputations as part of the anti-Shah movement, and they apparently were considered by Khomeini to be essential to the revolutionary movement (perhaps as a counterweight to the IRP and other radical elements). Furthermore, they were represented in the various institutions of the new revolutionary government, including the government cabinet, the Revolutionary Council, and later on the presidency under the new constitution.[29]

---

[28] For elaboration on this point see, in particular, Zabih (1982) and Taheri (1987).

[29] The weakness of these more secular elements lay in the fact that they had no mass political organization like the IRP. The Liberation Movement of Iran (and, indeed, the National Front and its allies) was little more than an elite collection of middle-class revolutionaries left over from the Mossadeq period. They had joined with the clerics and Khomeini, believing that strategy to be the only viable basis for generating mass support, and depended upon the leftist factions, both secular and religious, for disciplined organizational efforts among the youth and lower middle class. More so than the radical clergy, they were ultimately very dependent on the trust and support of Khomeini.

Hostage crisis politics point to a fourth actor in Iran's coalition decision unit on this issue: the student militants holding the American embassy. Never a unified group, the students involved in the embassy takeover ranged from leftist religious radicals who favored a relatively secular constitution to others with close ties to the radical clerics. Despite divisions on constitutional issues and other matters, the students shared the IRP's fear that the United States was planning to subvert the Iranian revolution and return the Shah to power, as it had allegedly done in 1953 (Roosevelt, 1979).[30] Their influence on the hostage issue lay in their immediate physical control over the lives of the Americans as well as the fact that they operated beyond the authority of even their supporters in the Revolutionary Council (Stempel, 1981; Zabih, 1982). Throughout the crisis, they successfully resisted efforts by moderate leaders to place the hostages under the control of the government, often with the support of radical leaders and occasionally even Khomeini.[31] But, at the same time, though, they were unable to carry out their demand that there should be public trials of the hostages if the Carter administration did not return the Shah and his alleged wealth to Iran.[32] The students holding the embassy did not appear willing to defy Khomeini, even though at the same time they were a wild card in the eyes of the government.

Taken together, then, policymaking within the Iranian government approached anarchy. Not only was power fragmented between contending political groups, but these groups were sharply polarized over basic questions regarding the future political order (linked to foreign policy) and, indeed, were competing for their very political survival. The near complete lack of any established decision rules compounded the situation to the point that it was often unclear just where power resided and which actors had the authority to act. The results were twofold. At one level, political infighting provoked a near continuous stream of

---

[30] On two occasions, in February and May 1979, members of the Fedayeen and Mujahidin (the leftist religious radicals) had attempted unsuccessfully to take possession of the American embassy, as a means of undercutting rapprochement with the West.

[31] Further complicating the situation was the ability of opposition groups (e.g., the Fedayeen and Mujahidin) outside the decision unit to put pressure on the students and the government, usually through IRP contacts, but occasionally in the streets, as did the Hesballahi.

[32] The decision to seize the embassy was apparently taken without the knowledge of the government, but certainly with at least the tacit approval of several of the radical clergy, some of whom came to the embassy the next day to tacitly and indirectly put the Ayatollah Khomeini's seal of approval on the operation. Within about a week the IRP leadership came to support the students' position, although they did not appear to have much enthusiasm for actually placing the hostages on trial.

anti-American and anti-Western criticism as various players demonstrated their nationalism credentials (and exposed any government attempts at pragmatic diplomacy to ease the crisis). At another level, an underlying deadlock persisted. No side was willing or able to alter the basic situation—that is, while the moderates were unable to gain the release (or simply control) of the hostages, the various radical elements were unsuccessful in their attempts to place the hostages on trial and punish them.

Despite all this, are there any factors that might have enabled or forced these contending factions to take substantively meaningful action to bring the situation to a close? One possibility is that the regime's "predominant actor" finally intervened to force a solution. In a way analogous to De Swaan's (1973) "pivotal actor," the predominant actor concept is not issue-specific but, instead, concerns the overall composition of the regime and the presence of one player who controls disproportionate amounts of key political resources (Achen, 1989; also Hagan, 1994). Clearly, Khomeini was predominant in this revolutionary regime. Although, as noted above, he was not actively involved in day-to-day affairs, he was a pivotal actor within a revolutionary coalition that could not directly challenge him. However, even though Khomeini surely had preferences on this issue, he withheld them at key points and instead saw this and other issues as ways of consolidating his power than as value preferences to be advocated. As noted above, his conception of leadership required that he be an arbiter of last resort, a philosophical guide rather than a strong executive. Hence, though his basic disposition was not to give the hostages back, he intervened rarely and only when the conflict among other members of the coalition became severe, for example, to reverse the March 1980 Revolutionary Council decision requiring the student militants to turn over the hostages to the Iranian government.

Even though there was no politically predominant actor in the regime willing to assert its influence, is there one actor with exclusive control over a critical resource who might exercise power? Druckman and Green's (1986) analysis suggests that such an actor may be able to independently implement certain policy options—in effect, operate as an undersized coalition. Especially in the situation of political anarchy, such an actor may well be willing to defy some members of the group and take extreme action—particularly if they believe they can mobilize wider public support to overrule the objections of other coalition players (who subsequently would not publicly oppose their action). The student militants were, arguably, in a position to act in this manner. They physically controlled the American hostages as well as the embassy and, as just noted, Khomeini did not permit the government to take over control of the embassy. They could have put the hostages on trial or worse. Of course, such did not occur. The student militants were apparently unwilling to defy the other members of the revolutionary coalition—or, more likely, were unwilling to risk defying Khomeini and his apparent wish to avoid more extreme punishment of the hostages. Whatever the dynamics, there was no member of the coalition able or willing to impose an agreement.

If no single actor is willing or able to prevail in such a highly fluid political setting, are there other factors that propel contending actors to cooperate? One possibility is a severe threat to the nation's security and/or the regime's political survival, a situation analogous to those that have led to "national unity cabinets" during wartime (e.g., Britain and Israel) as well as to Iran's response to the *foreign* threat posed by Iraq's invasion. In this type of case, an oversized coalition would appear to be the rational strategy for enforcing agreement, since no actor will want to risk incurring the wrath of other players (whose tolerance levels are not necessarily clear) by attempting to force through its own preferences. The oversized coalition allows members to isolate particular opponents entirely and/or, if necessary, attract uncommitted players to their position with side payments. Although such might have been the initial strategy of Iran's broad revolutionary coalition, as the hostage crisis proceeded, its membership became more and more restrictive. The hostage crisis issue, rather than unifying the country in response to U.S. pressure and international isolation, actually intensified political competition within the revolutionary coalition. There was no consensus on the extent of the American threat—while moderates worried about the cost of international isolation, radicals welcomed it as a means of purifying the revolution and breaking from the West. The hostage crisis placed the regime's legitimacy problem in sharp relief and enhanced rather than curtailed the domestic tensions within the revolution.

In sum, no factors helped overcome the basic deadlock within the Iranian government. None of the groups was willing or able to work together on the issue in a way that moved beyond simply continuing to hold the hostages in the student-controlled embassy. But, unlike the Dutch and Japanese cases previously discussed where rules existed and the situation was carefully contained, in the Iran case the deadlock was visible to all. The hostage crisis was marked by barrages of extreme anti-American rhetoric involving repeated threats as well as open criticisms of other members of the regime. Given the fluidity in the Iranian decision rules and the extreme distrust among the members of the revolutionary coalition, verbal pronouncements were often made by actors in order to openly undercut opponents. Decisions, in effect, took the form of fragmented symbolic action. Although verbal and contradictory, and in no way resolving the crisis, this ongoing verbal foreign policy was still significant. First, it politically undercut initiatives with the West to resolve the crisis and, second, it greatly inflamed tensions with the United States as the Carter administration took such rhetoric as indicative of Iranian intentions. Had rules existed and domestic conflict been contained (or, papered over as in the Dutch case), it is arguable that the crisis could have been handled more effectively—at least with respect to the international pressures Iran faced and the costs it ultimately paid.

The deadlock was broken—and then only gradually—when the political makeup of the regime changed. The period between the decision to keep the hostages and the final decision to release them was marked by intensifying international (including the Iraqi invasion) and U.S. pressures as well as major changes in the Iranian political scene. A presidential election held in February surprisingly was won by the moderate candidate, Bani Sadr, who viewed the hostage crisis as undercutting the revolution. His attempt in March 1980 to gain the release of the hostages was, however, reversed by a 7–6 vote in the Revolutionary Council when Khomeini, acting at the behest of the student militants and other fundamentalists, blocked a deal that would have transferred control of the hostages from the militant students to the Bani Sadr government (Salinger, 1981; Stempel, 1981:11). The radical clerics in the IRP had sided with the student militants because they saw resolution of the crisis as not only favoring the West but also shifting the internal power balance to Bani Sadr and the more moderate revolutionaries.

Only when the IRP had consolidated considerable power was it willing to tolerate negotiations. That came about as a result of the May election. IRP candidates acquired control of over two thirds of the Majles seats and promptly elected Hashemi-Rafsanjani to be its speaker, forcing Bani Sadr to appoint Mohammed Ali Rajai prime minister. The hard-line clerics were now politically dominant, and Beheshti, Speaker Rafsanjani, and Prime Minister Rajai were becoming a powerful triumvirate. Political infighting between Bani Sadr (who now had support from the leftists, including some revived elements of the People's Fedayeen) and the IRP continued with periodic fragmented verbal pronouncements in its foreign relations.[33] Yet emergence of a relatively coher-

---

[33] Although the political situation had changed considerably, the decision unit continued to be a coalition of autonomous actors. Khomeini remained as the predominant leader, but he still adhered to the role of the Faqih and avoided direct participation in governmental processes. He continued to delegate authority, not to any single individual, but to the Revolutionary Council, now increasingly united behind clerics of the IRP. It might be argued at this point that a single group—the Islamic Republican Party—had emerged as the single dominant actor within the government, if we grant that Beheshti, Rafsanjani, and Rajai compose such a single group. However, that assumption would ignore the political autonomy of the presidency under Bani Sadr as well as the militant students holding the hostages. Furthermore, under Bani Sadr, the presidency and executive branch returned to the role the PRG had played under Bazargan. The IRP and the clerics, indeed everyone on the Right, feared that Bani Sadr, an avowed secularist, would try to diminish clerical power. For that reason the surviving leftist People's Fedayeen and Mujahidin groups rallied around him, hence the presidency and those favorable to Bani Sadr's views became the opposite pole to the IRP. Both sides lobbied the student militants and tried to bring them around to their preferred policy conclusions.

ent government dominated by the IRP proved to be an important domestic political precondition to the eventual release of the hostages. In mid-September, Iran finally put forth an overture through the German embassy that led to the critical meeting signaling the onset of serious negotiations to end the hostage crisis.[34]

While the stabilization of the Iranian political scene reduced the domestic constraints surrounding the hostage issue, several unexpected events increased international pressure on the government to act to resolve the crisis. First, Iraq's attack on September 22 across a 400-mile front created intense pressure, especially on the radicals, to end the Western blockade of Iran and to obtain help. Second, Iran's isolation was further underlined in mid-October by Prime Minister Rajai's failure to get the U.N. General Assembly to condemn the Iraqi invasion. Third, the defeat of President Carter in his November 4 reelection bid, coupled with the much harder line taken by President-elect Reagan, shook the Iranians badly. As a result of these pressures, bargaining began in earnest, leading to the hostage release on the day of President Carter's departure from office.[35] Clearly the international situation had changed, but it should not be lost on the reader that the changes within Iran were even greater—the hardliners had finally been able to consolidate their power within relatively stable constitutional arrangements. Put more succinctly and in political terms, while the international costs of keeping the hostages had become more salient, the domestic benefits of keeping them had largely disappeared.

## SUMMARY AND CONCLUSIONS

The purpose of this article has been to sketch out a third type of decision unit—a coalition of multiple autonomous actors. Drawing upon theories of cabinet coalition formation, we have suggested a number of variables that govern the interactions of members of a coalition unit, that is, minimal size, policy

---

[34] A quiet, behind-the-scenes meeting was held in Bonn (Sept. 16–18) between a U.S. team led by Deputy Secretary of State Warren Christopher and an Iranian group led by Khomeini's son-in-law and confidante, Sadegh Tabatabai, a former deputy prime minister under Bazargan. In addition, on September 12, Khomeini finally announced his own conditions necessary for the return of the hostages: the U.S. was to pledge noninterference in Iranian affairs, return the Shah's money, unfreeze Iranian assets, and cancel all U.S. claims against the revolutionary government, including private ones.

[35] Accounts of the events leading to the release of the hostages are found in works by Stempel (1981), Zabih (1982), Sick (1985), and Bill (1988). Chapters by Robert Owen and Harold Saunders in Christopher (1985) give the best factual description of the diplomacy that took place during the hostage crisis.

space, pivotal actors, willingness and ability to bargain, and situational pressures at home and abroad. Furthermore, we have identified a key contingency variable—decision rules—that indicates coalitions may take one of three forms: (1) a "unit veto" model in which the coalition has established rules that require unanimity, (2) a "minimum connected winning coalition" model in which the coalition has established rules but they do not require unanimity, and (3) an "anarchy" model in which decision rules are largely absent. These three configurations have been explored, respectively, in the cases of the Netherlands and the 1979 cruise missile decision, Japan and the 1971 exchange rate crisis, and Iran and the 1979 hostage crisis.

The three cases examined here are by no means comprehensive, but they do suggest the diversity of decision unit structures and processes when the decision unit is a coalition—and the sharply divergent outcomes that can result. Two of the cases illustrate how extreme fragmentation in coalition decision units can lead to correspondingly extreme decisions—sustained political deadlock in two variations. The outcome of the Dutch handling of the cruise missile crisis was a very stable deadlock in which members of successive cabinets were unable to either accept or refuse the NATO missiles, while at the same time papering over their differences and arguably preventing the situation from exacerbating international and domestic tensions. The way Iran's revolutionary coalition handled the U.S. hostage crisis was much different. Although a deadlock persisted in which the hostages could neither be put on trial nor released, the torrent of verbal infighting targeted toward domestic and foreign audiences served to escalate domestic conflicts and international tensions. The Japanese case reminds us that coalitions do not necessarily produce extreme outcomes. One value of this latter case is that it shows how even a fragmented coalition government can produce a broad compromise in a reasonable amount of time.

It would be a mistake, though, to infer that the three types of coalition decision units portrayed here always manifest the outcomes evidenced in these three cases. Actually, the cruise missile, exchange rate, and hostage crises are arguably not "typical" of Dutch, Japanese, and, perhaps, even Iranian decision making. That is, coalitions in the Netherlands generally operate as a single group, the Japanese LDP often deadlocks, and the Iranian government did respond to the Iraqi invasion. A key point, in fact, in all three cases is that the respective decisions were due to interplay between other factors and coalition structure. The concept of the pivotal actor is relevant to all three settings, although not simply in the sense of imposing its own preferences. In the Dutch case, the Christian Democrats were pivotal in the sense of projecting their own incapacity to decide on the rest of the government, while in the Japanese case, the minister of finance acted cautiously in response to altered opinions. In the Iranian case, Khomeini did not take any position, but in doing so created a political vacuum that prolonged deadlock and prompted a larger political game among other actors in the coalition to seek legitimacy by asserting anti-Western nationalism.

Political beliefs and political relationships among coalition actors also interacted with coalition structure in helping to shape decisions. The relative positions (or policy space) among coalition members on the issues involved in the three cases were important to explaining what happened. The Japanese government was able to act, in part, because the MOF and BOJ were not polarized on the exchange rate issue, whereas the Dutch and Iranian debates reflected the more strongly held moderate and hard-line mentalities within each ruling circle. The willingness to bargain reinforced these tendencies. In the exchange rate crisis, Japanese leaders acted to keep the exchange rate issue from becoming politicized, while the Dutch leaders could not insulate their decision from the antimissile dissidents in the opposition and certain Iranian leaders had strong political incentives to openly politicize the hostage issue. These well-defined positions seem to have mediated the impact of the international pressures found in each of the cases. Only in the case of Japan did foreign threats and pressures reduce the level of disagreement in the coalition decision unit. In contrast, the intense NATO and American pressures faced by the Dutch and Iranian governments, respectively, propelled greater domestic alarm, intensifying not diminishing internal debate.

In addition to varying ability to achieve agreement (and avoid deadlock), the operation of these coalition decision units can also be examined in terms of "openness" to the environment, as raised in Margaret Hermann's theoretical overview in this special issue. The cases highlight differences in the extent to which coalitions are open to information from the political environment. Among the three cases, the Iranian coalition was clearly the most closed to any environmental pressures. The Iranian hard-liners were able to defy severe international pressure and to block pragmatic adjustments by the moderates, while domestically engineering the suppression of other political actors in the promotion of their domestic political agenda. More than in the other two cases, the Iranian decisions were driven by internal imperatives—namely, the competition for power among contending actors in the revolutionary coalition. The Dutch and Japanese decision units, in contrast, were relatively open to environmental signals as would be expected from their well-established decision rules. Such rules enabled coalition members to work together in coping with external pressures, but it does not mean that the two decision units responded in the same way to the pressures. The government of the Netherlands—like the Iranian government—defied severe international pressures. But the logic in the two governments was different. The Dutch government was unable to respond because it was severely constrained, not by internal dynamics, but by the wider domestic political environment, for example, Parliament and public opinion. The actions of the Japanese government reflected sensitivity to international pressures, mainly because LDP leaders kept the issue from becoming politicized publicly or drawn into internal LDP factional politics.

One further point is highlighted by these cases. Foreign policy episodes are usefully viewed as sequences of occasions for decisions that extend over time. The time frame may be a week or two, as in the Japan exchange rate crisis, or it may be a year or more—the Iranian hostage crisis lasted nearly a year and a half, while the Netherlands endured the cruise missile crisis for close to six years. In each of the cases described here, the governments made a number of decisions. Even the relatively responsive Japanese government first acted to resist the Nixon shocks and, only after "learning" the futility of protecting the yen, changed positions and decided to accept at least a partial (or dirty) float. In the Dutch case, as documented by Everts (1991), successive governments deadlocked. With regard to Iran, Stempel (1991) notes that there were several incomplete efforts at the release of the hostages. What is instructive about the latter two cases is not simply that they took longer. Rather, the deadlocks (in whatever form) were primarily the result of domestic political considerations—and ultimately led to the larger realignment of domestic actors. Dutch acceptance of the missiles was possible only after the demise of the dissidents with the decline of their public support and the weakening of their position in Parliament. Similarly, the Iranian government released the hostages after the hard-line clerics had established their dominance in that country's politics. These two cases illustrate that coalition decision making—even when deadlocked for prolonged periods—is not stagnant, but instead is dynamic like the other two types of decision units. The main difference, like so much about coalitions, is that foreign policymaking has to be seen within the larger domestic context.

## REFERENCES

ACHEN, CHRISTOPHER H. (1989) When Is a State with Bureaucratic Politics Representable as a Unitary Rational Actor? Paper presented at the annual meeting of the International Studies Association, London, March–April.

ANDREWS, WILLIAM G. (1962) *French Politics and Algeria: The Process of Policy Formation, 1954–1962.* New York: Appleton-Century-Crofts.

ASPATURIAN, VERNON (1966) "Internal Politics and Foreign Policy in the Soviet System." In *Approaches to Comparative and International Politics*, edited by R. Barry Farrell. Evanston, IL: Northwestern University Press.

AXELROD, ROBERT (1970) *Conflict of Interest.* Chicago: Markham.

BALDWIN, DAVID A. (1993) *Neorealism and Neoliberalism: The Contemporary Debate.* New York: Columbia University Press.

BARNETT, A. DOAK (1985) *The Making of Foreign Policy in China: Structure and Process.* Boulder, CO: Westview Press.

BAR-SIMON-TOV, YAACOV (1983) *Linkage Politics in the Middle East: Syria Between Domestic and External Conflict, 1961–1970.* Boulder, CO: Westview Press.

BAYLIS, THOMAS (1989) *Governing by Committee: Collegial Leadership in Advanced Societies.* Albany: State University of New York Press.

BILL, JAMES A. (1988) *The Eagle and the Lion: The Tragedy of American-Iranian Relations.* New Haven, CT: Yale University Press.

BOGDANOR, VERNON (1983) *Coalition Government in Western Europe.* London: Heinemann Educational Books.

BROWNE, ERIC C., AND JOHN DREIJMANIS (1982) *Government Coalitions in Western Democracies.* New York: Longman Press.

BROWNE, ERIC C., AND JOHN P. FRENDREIS (1980) Allocating Coalition Payoffs by Conventional Norm: An Assessment of the Evidence from Cabinet Coalition Situations. *American Journal of Political Science* **24**(4):753–768.

CARALEY, DEMETRIOS (1966) *The Politics of Military Unification.* New York: Columbia University Press.

CHRISTOPHER, WARREN (1985) *American Hostages in Iran: The Conduct of a Crisis.* New Haven, CT: Yale University Press.

DAVIS, VINCENT (1967) *The Admirals Lobby.* Chapel Hill: University of North Carolina Press.

DESTLER, I. M. (1980) *Making Foreign Economic Policy.* Washington, DC: Brookings Institution Press.

DESTLER, I. M. (1986) *American Trade Politics: System Under Stress.* New York: Twentieth Century Fund.

DESTLER, I. M., HARUHIRO FUKUI, AND HIDEO SATO (1979) *The Textile Wrangle: Conflict in Japanese-American Relations, 1969–1971.* Ithaca, NY: Cornell University Press.

DESTLER, I. M., LESLIE H. GELB, AND ANTHONY LAKE (1984) *Our Own Worst Enemy: The Unmaking of American Foreign Policy.* New York: Simon and Schuster.

DESTLER, I. M., PRISCILLA CLAPP, HIDEO SATO, AND HARUHIRO FUKUI (1976) *Managing an Alliance: The Politics of U.S.-Japanese Relations.* Washington, DC: Brookings Institution.

DE SWAAN, ABRAM (1973) *Coalition Theories and Cabinet Formations.* Amsterdam: Elsevier.

DODD, LAWRENCE (1976) *Coalitions in Parliamentary Government*. Princeton, NJ: Princeton University Press.

DRUCKMAN, DANIEL, AND JUSTIN J. GREEN (1986) *Political Stability in the Philippines*. University of Denver Monograph Series.

EENENNAAM, B. J. VAN (1988) *Achtenveertig kruisraketten. Hoogspanning in de Lage landen*. The Hague: Staatsuitgeverij.

EVERTS, PHILIP P. (1983) *Public Opinion, the Churches, and Foreign Policy: Studies of Domestic Factors in the Making of Dutch Foreign Policy*. Leiden: Institute for International Studies.

EVERTS, PHILIP P. (1985) *Controversies at Home: Domestic Factors in the Foreign Policy of the Netherlands*. Boston: Martinus Nijhoff.

EVERTS, PHILIP P. (1991) Between the Devil and the Deep Blue Sea: 48 Cruise Missiles for the Netherlands. Paper presented at the conference on How Decision Units Shape Foreign Policy, Jackson Hole, Wyoming, November 1989.

FRANK, THOMAS M., AND EDWARD WEISBAND (1979) *Foreign Policy by Congress*. New York: Oxford University Press.

FUKUI, HARUHIRO (1970) *Party in Power: The Japanese Liberal-Democrats and Policy Making*. Berkeley: University of California Press.

FUKUI, HARUHIRO (1975) Okinawa henkan kosho: Nihon seifu ni okeru kettei katei [The Okinawa Reversion Negotiations: Decision Making in the Japanese Government]. *Kokusai seiji* (May):97–124.

FUKUI, HARUHIRO (1977a) Foreign Policy Making by Improvisation: The Japanese Experience. *International Journal* **32**:791–812.

FUKUI, HARUHIRO (1977b) "Tanaka Goes to Peking: A Case Study in Foreign Policy Making." In *Policymaking in Contemporary Japan*, edited by T. J. Pempel. Ithaca, NY: Cornell University Press.

FUKUI, HARUHIRO (1979) "The GATT Tokyo Round: The Bureaucratic Politics of Multilateral Diplomacy." In *The Politics of Trade: U.S. and Japanese Policymaking for the GATT Negotiations*, edited by Michael Baker. New York: Occasional Papers of the East Asian Institute.

FUKUI, HARUHIRO (1987) Too Many Captains in Japan's Internationalization: Travails at the Foreign Ministry. *Journal of Japanese Studies* (summer): 359–382.

FUKUI, HARUHIRO (1989) Japanese Decision Making in the 1971 Exchange Rate Crisis. Paper presented at the conference on How Decisions Units Shape Foreign Policy, Jackson Hole, Wyoming, November.

GELMAN, HARRY (1984) *The Brezhnev Politburo and the Decline of Detente*. Ithaca, NY: Cornell University Press.

GOLDMANN, KJELL, STEN BERGLUND, AND GUNNAR SJOSTEDT (1986) *Democracy and Foreign Policy: The Case of Sweden*. Brookfield, VT: Gower.

HAGAN, JOE D. (1993) *Political Opposition and Foreign Policy in Comparative Perspective*. Boulder, CO: Lynne Rienner.

HAGAN, JOE D. (1994) Domestic Political Systems and War Proneness. *Mershon International Studies Review* **38**:183–207.

HAMMOND, PAUL Y. (1963) "Super Carriers and B-36 Bombers: Appropriations, Strategy and Politics." In *American Civil-Military Decisions: A Book of Case Studies*, edited by Harold Stein. University: University of Alabama Press.

HANRIEDER, WOLFRAM F. (1970) *The Stable Crisis: Two Decades of German Foreign Policy*. New York: Harper and Row.

HANRIEDER, WOLFRAM F., AND GRAEME P. AUTON (1980) *The Foreign Policies of West Germany, France, and Britain*. Englewood Cliffs, NJ: Prentice-Hall.

HELLMANN, DONALD (1969) *Japanese Foreign Policy and Domestic Politics: The Peace Agreement with the Soviet Union*. Berkeley: University of California Press.

HINCKLEY, BARBARA (1981) *Coalitions and Politics*. New York: Harcourt Brace Jovanovich.

HINTON, HAROLD C. (1972) *China's Turbulent Quest: An Analysis of China's Foreign Policy Since 1949*. New York: Macmillan.

HOSOYA, CHIHIRO (1976) "Japan's Decision-Making System as a Determining Factor in Japan-United States Relations." In *Japan, America, and the Future World Order*, edited by Morton A. Kaplan and Kinhide Mushakoji. Chicago: Free Press.

HUNTINGTON, SAMUEL P. (1968) *Political Order in Changing Societies*. New Haven, CT: Yale University Press.

JOHNSON, CHALMERS (1982) *MITI and the Japanese Miracle: The Growth of Industrial Policy*. Stanford, CA: Stanford University Press.

KAARBO, JULIET (1996) Power and Influence in Foreign Policy Decision Making: The Role of Junior Coalition Partners in German and Israeli Foreign Policy. *International Studies Quarterly* **40**:501–530.

KAPLAN, MORTON A. (1957) *System and Process in International Politics*. New York: John Wiley.

KEGLEY, CHARLES W., JR. (1987) "Decision Regimes and the Comparative Study of Foreign Policy." In *New Directions in the Study of Foreign Policy*, edited by Charles F. Hermann, Charles W. Kegley, Jr., and James N. Rosenau. Boston: Allen and Unwin.

KEOHANE, ROBERT O. (1984) *After Hegemony: Cooperation and Discord in the World Political Economy*. Princeton, NJ: Princeton University Press.

KEOHANE, ROBERT O. (1989) *International Institutions and State Power*. Boulder, CO: Westview Press.

KEOHANE, ROBERT O., JOSEPH S. NYE, AND STANLEY HOFFMANN (1993) *After the Cold War: International Institutions and State Strategies in Europe, 1989–1991*. Cambridge, MA: Harvard University Press.

KRAUSS, ELLIS S., THOMAS P. ROHLEN, AND PATRICIA G. STEINHOFF (1984) *Conflict in Japan*. Honolulu: University of Hawaii.

KUSUDA, MINORU (1983) *Sato seiken 2797 nichi* [The Sato government's 2797 days], 2 vols. Tokyo: Gyosei mondai kenkyujo.

LALMAN, DAVID, JOE OPPENHEIMER, AND PIOTR SWISTAK (1993) "Formal Rational Choice Theory: A Cumulative Science of Politics." In *The State of the Discipline II*, edited by Ada Finifter. Washington, DC: American Political Science Association.

LAMBORN, ALAN C. (1991) *The Price of Power: Risk and Foreign Policy in Britain, France, and Germany*. Boston: Allen and Unwin.

LEISERSON, MICHAEL (1966) Coalitions in Politics. Ph.D. dissertation, Yale University.

LEPPER, MARY M. (1971) *Foreign Policy Formulation: A Case Study of the Nuclear Test Ban Treaty of 1963*. Columbus, OH: Merrill.

LEPRESTRE, PHILIPPE G. (1984) "Lessons of Cohabitation." In *French Security in a Disarming World*, edited by Philippe G. LePrestre. Boulder, CO: Lynne Rienner.

LEVY, JACK S., AND LILY VAKILI (1990) External Scapegoating by Authoritarian Regimes: Argentina in the Falklands/Malvinas Case. Manuscript, Rutgers University.

LIJPHART, AREND (1968) *The Politics of Accommodation: Pluralism and Democracy in the Netherlands*. Berkeley: University of California Press.

LIJPHART, AREND (1984) *Democracies: Patterns of Majoritarian and Consensus Government in Twenty-One Countries*. New Haven, CT: Yale University Press.

LINCOLN, JENNIE K., AND ELIZABETH G. FERRIS (1984) *The Dynamics of Latin American Foreign Policies: Challenges for the 1980s*. Boulder, CO: Westview Press.

LINDEN, CARL A. (1978) *Khrushchev and the Soviet Leadership, 1957–1964*. Baltimore, MD: Johns Hopkins University Press.

LINDSAY, JAMES M. (1994) Congress, Foreign Policy, and the New Institutionalism. *International Studies Quarterly* **38**(2):281–304.

LUEBBERT, GREGORY (1984) A Theory of Government Formation. *Comparative Political Studies* **17**(2):229–264.

LUEBBERT, GREGORY (1986) *Comparative Democracy: Policymaking and Governing Coalitions in Europe and Israel.* New York: Columbia University Press.

MANSFIELD, EDWARD D., AND JACK SNYDER (1995) Democratization and the Danger of War. *International Security* **20**(1):5–38.

MORSE, EDWARD L. (1973) *Foreign Policy and Interdependence in Gaullist France.* Princeton, NJ: Princeton University Press.

MUNOZ, HERALDO, AND JOSEPH S. TULCHIN (1984) *Latin American Nations in World Politics.* Boulder, CO: Westview Press.

OGATA, SADAKO (1988) *Normalization with China: A Comparative Study of U.S. and Japanese Process.* Berkeley, CA: Institute for East Asian Studies.

OKAMOTO, FUMIO (1972) *Sato seiken* [The Sato Government]. Tokyo: Hakubasha.

OKURASHO DAIJIN KAMBO CHOSA KIKAKUSHITSU [Ministry of Finance, Minister's Secretariat, Office of Research and Planning] (1982) Showa 30-nendai iko no zaisei kinvu seisaku no ashidori [The Record of Fiscal and Monetary Policies Since the Mid-1950s]. Tokyo: Zaisei shohosha.

ORI, KAN (1976) "Political Factors in Postwar Japan's Foreign Policy Decisions." In *Japan, America, and the Future World Order*, edited by Morton A. Kaplan and Kinhide Mushakoji. New York: Free Press.

PEMPEL, T. J. (1982) *Policy and Politics in Japan: Creative Conservatism.* Philadelphia: Temple University Press.

PERLMUTTER, AMOS (1981) *Modern Authoritarianism: A Comparative Institutional Analysis.* New Haven, CT: Yale University Press.

PETERSON, SUSAN (1996) *Crisis Bargaining and the State: The Domestic Politics of International Conflict.* Ann Arbor: University of Michigan Press.

PRIDHAM, GEOFFREY (1986) *Coalitional Behaviour in Theory and Practice.* London: Cambridge University Press.

PUTNAM, ROBERT (1988) Diplomacy and Domestic Politics: The Logic of Two-Level Games. *International Organization* **42**(3):427–460.

RAMAZANI, R. K. (1990) *Iran's Revolution: The Search for Consensus.* Bloomington: Indiana University Press.

RICHARDSON, BRADLEY M., AND SCOTT C. FLANAGAN (1984) *Politics in Japan.* Boston: Little, Brown.

RIKER, WILLIAM (1962) *The Theory of Political Coalitions*. New Haven, CT: Yale University Press.

RISSE-KAPPEN, THOMAS (1991) Public Opinion, Domestic Structure, and Foreign Policy in Liberal Democracies. *World Politics* **43**:479–512.

ROOSEVELT, K. (1979) *Countercoup: The Struggle for Control of Iran*. New York: McGraw-Hill.

ROSECRANCE, RICHARD, AND ARTHUR A. STEIN (1993) *The Domestic Bases of Grand Strategy*. Ithaca, NY: Cornell University Press.

SALINGER, PIERRE (1981) *America Held Hostage: The Secret Negotiations*. Garden City, NY: Doubleday.

SASAKI, TAKAO (1973) Infure to shotoku seisaku [Inflation and Income Policy]. Kokumin seiji kenkvukai getsuvokai repoto [National Politics Study Association, Monday Meeting Reports], October 8. Tokyo: Kokumin seiji kenkyukai.

SCHILLING, WARNER R., PAUL Y. HAMMOND, AND GLENN H. SNYDER (1962) *Strategy, Politics, and Defense Budgets*. New York: Columbia University Press.

SHINOHARA, MIYOHEI (1961) *Nihon keizai no seicho to junkan* [The Growth and Cycle of the Japanese Economy]. Tokyo: Sobunsha.

SHINOHARA, MIYOHEI (1973) Kawase reto to sengo keizai seicho [The Exchange Rate and Postwar Economic Growth]. In *Tenki ni tatsu nihon keizai* [The Japanese Economy at a Cross-Roads], edited by Keizai tembo kondankai. Tokyo: Tokyo daigaku shuppankai.

SICK, GARY (1985) *All Fall Down*. New York: Random House.

SMITH, MICHAEL, STEVE SMITH, AND BRIAN WHITE (1988) *British Foreign Policy: Tradition, Change, and Transformation*. London: Unwin and Hyman.

SNYDER, GLENN H., AND PAUL DIESING (1977) *Conflict Among Nations: Bargaining, Decision Making, and System Structure in International Crises*. Princeton, NJ: Princeton University Press.

SNYDER, JACK (1991) *Myths of Empire: Domestic Policies and International Ambition*. Ithaca, NY: Cornell University Press.

SOETENDORP, R. B. (1989) "The NATO 'Doubletrack' Decision of 1979." In *The Politics of Persuasion: Implementation of Foreign Policy by the Netherlands*, edited by Philip P. Everts and G. Walraven. Aldershot, England: Avebury.

SPANIER, JOHN, AND JOSEPH NOGEE (1981) *Congress, the Presidency, and American Foreign Policy*. New York: Pergamon.

STADEN, A. VAN (1985) "To Deploy or Not to Deploy: The Case of the Cruise Missiles." In *Controversies at Home: Domestic Factors in the Making of Foreign Policy*, edited by Philip P. Everts. Dordrecht: Martinus Nijhoff.

STEINER, JURG (1974) *Amicable Agreement Versus Majority Rule*. Chapel Hill: University of North Carolina Press.

STEMPEL, JOHN D. (1981) *Inside the Iranian Revolution*. Bloomington: Indiana University Press.

STEMPEL, JOHN D. (1989) The Iranian Hostage Crisis: Non-Decision by Revolutionary Coalition. Paper presented at the conference on How Decision Units Shape Foreign Policy, Jackson Hole, Wyoming, November.

SUNDELIUS, BENGT (1982) *Foreign Policies of Northern Europe*. Boulder, CO: Westview Press.

TAHERI, A. (1987) *Hold Terror: The Inside Story of Islamic Terrorism*. London: Sphere Books.

TAKEUCHI, KATSUNOBU (1988) *Nempvo de miru nihon keizai no ashidori* [The Footsteps of the Japanese Economy as Seen in a Chronicle]. Tokyo: Zaisei shohosha.

THELEN, KATHLEEN, AND SVEN STEINMO (1992) "Historical Institutionalism in Comparative Politics." In *Structuring Politics: Historical Institutionalism in Comparative Analysis*, edited by Sven Steinmo, Kathleen Thelen, and Frank Longstreth. Cambridge: Cambridge University Press.

TSEBELIS, GEORGE (1990) *Nested Games: Rational Choice in Comparative Politics*. Berkeley: University of California Press.

UCHINO, TATSUO (1976) "Nikuson shokku: l-doru 308 yen ni [The Nixon Shock: $1 Becomes 308 Yen]." In *Showa keizaishi* [The Economic History of the Showa Era], edited by Arisawa Hiromi. Tokyo: Nihon keizai shimbunsha.

VALENTA, JIRI (1979) *Soviet Intervention in Czechoslovakia, 1968: Anatomy of a Decision*. Baltimore, MD: Johns Hopkins University Press.

VASQUEZ, JOHN A. (1993) *The War Puzzle*. New York: Cambridge University Press.

VERNON, RAYMOND, DEBORAH L. SPAR, AND GREGORY TOBIN (1991) *Iron Triangles and Revolving Doors: Cases in U.S. Foreign Economic Policymaking*. New York: Praeger.

VOGEL, EZRA (1979) *Japan as Number One*. Cambridge, MA: Harvard University Press.

WALT, STEPHEN (1996) *Revolution and War*. Princeton, NJ: Princeton University Press.

WATANABE, AKIO (1981) Dai-63-dai: Dai-3-ji Sato naikaku: Gekido no 70-nendai e hashiwatashi [63rd: the 3rd Sato Cabinet: Bridge-Building for Passage to the Turbulent 1970s]. In *Nihon naikaku shiroku* [Historical Records of Japanese Cabinets], vol. 6, edited by Hayashi Shigeru and Tsuji Kiyoaki. Tokyo: Daiichi hoki shuppan.

WEINSTEIN, FRANKLIN B. (1976) *Indonesian Foreign Policy and the Dilemma of Dependence: From Sukarno to Soeharto*. Ithaca, NY: Cornell University Press.

WELFIELD, JOHN (1976) "Japan, the United States, and China in the Last Decade of the Cold War: An Interpretive Essay." In *The International Yearbook of Foreign Policy Analysis*, vol. 2, edited by Peter Jones. London: Croom and Helm.

YAMAMURA, YOSHIHARU (1984) *Sengo nihon gaikoshi* [The Diplomatic History of Postwar Japan], vol. 5. Tokyo: Sanseido.

YOSHITOMI, MASARU (1977) *Gendai nihon keizai shiron* [The Contemporary Japanese Economy]. Tokyo: Toyo keizai shimposha.

ZABIH, S. (1982) *Iran Since the Revolution*. Washington, DC: Johns Hopkins University Press.

# People and Processes in Foreign Policymaking:

## Insights from Comparative Case Studies

*Ryan K. Beasley*
Baker University

*Juliet Kaarbo*
University of Kansas

*Charles F. Hermann*
Bush School, Texas A & M University

*Margaret G. Hermann*
Maxwell School, Syracuse University

The previous articles in this special issue have elaborated a framework for classifying the people involved in foreign policymaking into decision units. Of particular interest has been examining the circumstances under which one type of decision unit takes responsibility for making the choice regarding how to deal with a foreign policy problem and the effect of the nature of that decision unit on the substance of the action selected. The present article is intended to report the results of the application of the framework to sixty-five case studies involving foreign policy issues facing thirty-one countries from all regions of the world. A list of the cases can be found in the appendix.

In applying the framework, specialists and advanced graduate students were asked to focus on a particular foreign policy problem, identify specific occasions for decision that confronted the government in dealing with the problem, decide which of the three types of decision units had the authority to commit the resources of the government for a specific occasion for decision, and examine the decision process leading to the selection of a foreign policy action regarding that occasion for decision. Beyond building a cross-national set of case studies that employ the same framework, we were interested in the answers to three questions. (1) How easy is it to apply the framework to actual cases of foreign policymaking; can the queries in the framework be answered with the kinds of case materials that are available? (2) Do the variables and theories embedded in the framework correspond at all with the factors that historical evidence suggests are central to conceptualizing the decision process? (3) What lessons can we learn from the case applications to help us elaborate and refine the framework further?

We make no claim that the set of cases reported on here are a representative sample. A number of the cases (23 percent) were selected because the individuals knowledgeable about the case materials were also scholars with a strong interest in improving the conceptual rigor in comparative foreign policy analysis and had already done extensive research on the case to be examined. We were not asking for new research but rather for an application of the framework to ground familiar to each. The cases used as illustrations in the discussions of the three types of decision units in the previous articles were completed by these specialists. We asked them to consider how the model could be made richer theoretically to capture what happened in the historical situations. The rest of the cases (77 percent) were done by advanced graduate students who were interested in studying particular countries and governments; they wanted to increase their expertise regarding a region of the world and to explore the ramifications of the framework for understanding how foreign policy is made.[1] In their case studies, the students indicated their degree of confidence in the choices they were making in applying the framework given the materials available to them.

An attempt was made to include cases that showcased each of the three types of decision units. Some 40 percent (26) of the cases focused on decision units that involved predominant leaders, 37 percent (24) on decision units that were single groups, and 23 percent (15) on decision units that were coalitions.

---

[1]The graduate students who did the case studies were participants in several summer workshops devoted to exploring the decision units framework at the Mershon Center, Ohio State University, as well as students in Comparative Foreign Policy Analysis courses at Ohio State University and in the Maxwell School, Syracuse University, taught by Margaret Hermann.

The countries represented in the case studies ranged in their degree of democratization, their level of development, and their status in the international system. We were also interested in exploring a mix of types of issues. Thus, the cases examined decisions to use force in response to immediate military concerns, attempts to negotiate conflicts peacefully, economic interactions, initiation of diplomatic activities, and problems growing out of human rights and environmental concerns. Although in a majority of the cases the governments were addressing time-urgent situations, roughly one quarter of the cases were more routine in nature. All but one of the cases occurred during the post-World War II era.

In what follows, we will report what we have learned about the decision units framework from the case studies and from the case analysts' comments after applying the framework to case materials. While none of the analysts found it impossible to use the framework, all provided us with insights about how it did, could, or should work. In several instances, these insights raised issues we had not considered before; in other instances, they helped to clarify problems we had identified but not yet resolved. Overall the analysts' questions posed challenges to the framework in three areas: (1) the overall importance of a decision-making perspective; (2) the difficulty in making some of the conceptual distinctions demanded by the framework; and (3) the comprehensiveness of the framework. Let us examine each of these areas in more detail.

## ARE DECISION UNITS IMPORTANT IN SHAPING FOREIGN POLICY BEHAVIOR?

A basic premise of the theoretical effort described in this special issue is that the foreign policy actions of governments are shaped in significant ways by the nature of the unit involved in the decision-making process. Although we recognize that numerous domestic and international factors can, and do, influence foreign policy activity, we argue that these influences are channeled through the political structure of a government that identifies, decides, and implements foreign policy. We propose that the configuration and dynamics of the decision unit can affect the foreign policy action that is chosen, particularly if the resulting decision unit has the ability both to commit the resources of the society and to make a decision that cannot be readily reversed. In effect, the decision unit perceives and interprets the pressures and constraints posed by the domestic and international environments. Do the sixty-five case studies lend support to our premise? In other words, do the decisions specified by the framework match the choices made in the historical cases? And, are there any patterns to the effects that the decision units seem to have on governments' foreign policy behavior?

## Match Between Framework-Determined and Historical Decisions

An examination of the match between the decisions resulting from the application of the framework with what happened in the historical cases indicated an overlap in fifty-four of the sixty-five cases (83 percent).[2] Case analysts were asked to compare the process outcomes they identified through applying the framework to their cases with what the historical material suggested happened. The authors determined the extent of the match based on the case analysts' comparisons. Recall from the previous description of the decision units framework that process outcomes denote whose positions have counted in the final decision and indicate the end point of the decision-making process. There are six possible outcomes: concurrence, one party's position prevails, mutual compromise/consensus, lopsided compromise, deadlock, and fragmented symbolic action.

The advanced graduate student case analysts also indicated their confidence in the choices they made in applying the decision units framework. They did so by noting high and low confidence in their decisions at the various choice points posed by the framework. Since the students generally had to rely on secondary sources for information on their cases, this indicator also reflects the amount of information that was available on which to make the determinations required by the framework. On average, the student analysts reported high confidence in their choices 75 percent of the time. For the cases where there was a match between the predicted outcome using the framework and the historical outcome, average high confidence was 85 percent; for the cases with a poor fit, average high confidence was 45 percent. More revealing, perhaps, are the data that show 95 percent of the cases where all the ratings of the analysts indicated high confidence exhibited an overlap between the outcomes proposed by applying the framework and reported in the case material. This percentage was 47 for those cases where none of the ratings of the analysts were high in confidence. Analysts indicated, on average, higher confidence in their choices in applying the framework to cases involving a predominant leader decision unit (80 percent) than the other two types of decision units. The average percentage of high confidence ratings for cases with a single group decision unit was 70 percent and for cases with a coalition decision unit, 67 percent. These data suggest that when there was sufficient detail available on how a particular decision was made the analysts were confident in the choices they were asked to make via the framework. Examining the process provided information about the resulting decision that was a good fit to what appeared to actually happen.

---

[2] The data on which this analysis and the others reported in this article are based are available from the authors at mgherman@maxwell.syr.edu.

An examination of the "misses"—those cases where the framework failed to lead to a choice that resembled history—permits several observations about the relevance of decision units to understanding governments' foreign policy actions. The framework provides little extra information about a decision when all members of the decision unit concur on what should be done from the outset (or, if the decision unit is a predominant leader, the leader has a position that is reinforced by outside information). Consider, for example, the British cabinet's response to the invasion of the Falkland Islands or the Greek government's response to the Turkish intervention into Cyprus following the ouster of President Makarios. In cases like these where there is a foreign threat or challenge to the legitimacy of the government, members of the decision unit tend to put aside disagreements among themselves and focus their attention on the survival of the regime. The decision-making process is short-circuited as the members of the decision unit share a position regarding what needs to be done from the outset and act on this position.

The decision units framework described here becomes important when the members of the decision unit disagree or, at the least, do not have an initial consensus about what should be done in response to a particular occasion for decision and problem. In such instances, the framework provides one way of understanding how such decision units are likely to cope with these differences based on how they are configured. Of the sixty-five cases we have explored, 71 percent involved decision units where there was dissensus among those involved in the policymaking process. Often until these disagreements are resolved and some kind of consensus is built, little action is possible. The framework facilitates our saying whose positions are likely to count in this type of policymaking setting and indicates the process likely to be favored as well as what decision will probably be made.

The decision units framework also becomes useful in deciphering anomalous situations where the structure of the political system suggests one kind of policymaking is in effect while the domestic politics of the moment dictate another. Consider the incident where U.S. hostages were taken in Iran in 1979. American policymakers believed they were dealing with a regime that had a predominant leader, the Ayatollah Khomeini. Application of the decision units framework indicates, instead, that they were dealing with a coalition of actors in the process of determining the nature of the Iranian political structure following the revolution (see discussion of this case in the previous article on the coalition decision unit). The many variations in the behavior of the Iranian government in dealing with the problem was more reflective of the fragmented symbolic action often characteristic of coalition decision units where the various actors cannot agree than of the actions of a decision unit headed by a predominant leader. A more contemporary example involves the current Chinese leadership. Following in the footsteps of a Mao Zedong and Deng Xiaoping, it is an easy assumption to make that Jiang Zemin, like these other two, is a predominant leader. But examinations of the negotiations between the United States and China over the latter's entry into the World Trade Organization in the spring of 1999 and the Chinese government's reaction to the NATO bombing of their Belgrade embassy in May 1999 using the decision units framework reveal that these decisions were made by a single group composed of individuals with positions ranging from hard-line to moderate regarding the manner in which the Chinese government should deal with the United States (Chen and Chung, 2000; see also Gilboy and Heginbotham, 2001).

## *Effects on Foreign Policy Behavior*

An examination of the sixty-five cases suggests that what goes on in the decision unit can affect governments' foreign policy behavior by highlighting or reinforcing domestic, international, or cultural constraints and pressures and, thus, amplifying the importance of such constraints on what happens, or by diminishing such constraints and pressures by giving them less credence. In the first instance, the decision unit reduces its own effect on foreign policy while in the second it enhances its role. Asking graduate students in a methods course to indicate which of the outcomes in the sixty-five cases appeared to emphasize the nature of the political context and which the nature of the decision unit revealed that 51 percent of the time the configuration of people making the authoritative decision had a greater effect on the choice than the constraints in the environment; 46 percent of the time what was happening in the political landscape was highlighted by the decision. The other 4 percent could not be classified.[3] Let us review some of the case examples from previous articles in this special issue to illustrate both types of decision units.

In analyzing Dutch decision making with regard to NATO's proposed deployment of cruise missiles as advocated by the Reagan administration, it was observed that forces in the society that both favored and opposed deployment were represented in the coalition cabinets charged with making a decision. The debate being waged in the domestic environment outside the cabinet was reinforced in the policymaking process as an antideployment wing of one of the major coalition parties prevented their leaders in the cabinet from supporting other party leaders and foreign ministry officials in accepting the missiles. The upshot of the division was a deadlock until the domestic forces against deployment calmed down.

International and cultural constraints can also be reinforced by the decision unit. In examining the Israeli government's decisions in 1967 in response to a

---

[3] The inter-coder agreement among the students for these judgments averaged .88. Twelve students participated in the exercise.

clear threat from the Arab states, we note that policymakers sought information about U.S. support for military action before committing themselves to an action. Knowledge about what the United States would condone was important to shaping the options the Israelis believed they had in this particular instance. The importance of the norm of neutrality for the Swedes affected their government's response to the Soviet submarine stuck in their waters. Conditioned to live by this norm, Swedish policymakers did not consider alternatives that had any chance of jeopardizing neutrality.

But decision units also diminish the relevance of outside constraints, paying little attention to, or even dismissing, external pressures. These units shape what the government does more directly. Consider the Nigerian government's quick recognition of the MPLA as the legitimate government in Angola in 1975. This response can be traced to the presence of a strong political leader, Murtala Mohammed, who was intensely interested in this issue and in building an activist foreign policy for Nigeria as well as improving his country's power and legitimacy in inter-African affairs. The leadership style of a predominant leader overrode a wide variety of constraints that might have precluded such strong action by a pro-Western African regime. An examination of the British decisions during the Munich crisis also show the importance of what goes on in the decision unit in shaping what happens. Throughout the crisis, the inner cabinet included a leader who had strong preferences and members who raised questions about these preferences. The power, respect, and status of these opponents vis-à-vis the leader at any point in time had a greater impact on the decision than did information about the situation. Similarly, in the Iranian response to the student takeover of the U.S. embassy in Teheran in 1979, the struggle for power between the more militant mullahs, the moderates, the student captors, and the Ayatollah Khomeini coalesced around the hostages with all parties needing to participate in any decision to maintain their credibility as a force to be reckoned with in domestic politics. Who would win depended on who could exert the most influence in the decision unit on what happened with the hostages.

Whether the decision unit is likely to amplify or diminish the influence of external constraints appears to depend on the nature of the decision unit itself, in particular, the nature of the key contingencies in the parlance of the framework being discussed here. Decision units that amplify external constraints tend to be those that are more open to outside influences such as more sensitive predominant leaders, single groups with limited loyalty to the group itself and a unanimity decision rule, and coalitions where unanimity is required for a decision to be acceptable. In each of these types of decision units, information about the domestic and international environments is important to shaping what the policymakers see as feasible alternatives as well as to helping them define the problem they are facing. The decision unit is structured so that policymakers are interested in seeking information from outside the unit to understand where important constituencies stand and to determine the options they can support. The members of the decision unit perceive a certain responsibility to these external forces and bring them into the decision process.

Decision units that tend to diminish the effects of outside constraints seem generally more closed to what is going on around them. The focus of attention is on what is happening inside the decision unit—on what the relatively insensitive predominant leader wants, on maintaining the cohesion of the group for members whose identity resides in the group itself, on solving the problem for members of the single group with loyalty elsewhere but with a majority decision rule, on winning out over the others for members of the coalition with no rules, and on reaching some modicum of agreement among members of the coalition with rules that allow for majority rule. The members of these decision units are turned inward and are intent on achieving a particular goal. They become rather oblivious to external forces; outside information intrudes only as it helps to bolster a member's position.

In another place, two of the authors (Hermann and Hermann, 1989) found a relationship between this open-closed phenomenon and governments' foreign policy behavior. In an examination of the foreign policy activity of twenty-five countries, the closed decision units (those that tend to diminish external pressures) engaged in more extreme foreign policy activity than the open decision units (those that tend to amplify external constraints). The open units displayed more cautious behavior, resorting to compromise more than the closed units. Indeed, the closed units were more likely to push their positions, to commit their resources, and to be conflictual in the foreign policy arena than the open units. In essence, the open units were more constrained by the situation in their foreign policy behavior while the closed units appeared to go their own way. Hagan (1993) found a similar set of results examining how these different types of decision units related to domestic opposition. The open units were more likely to accommodate to opposition; the closed units to challenge (suppress, scapegoat, highlight differences, co-opt, etc.) any opposition.

This analysis seems to suggest that an understanding of how decision units shape foreign policy may be more relevant if the decision units being studied have the characteristics of a closed rather than an open unit. But, on reflection, an examination of the open units provides the analyst with information about which external forces are probably going to be taken into consideration by the decision unit in its deliberations and how such constraints are likely to affect the decision. Moreover, if the decision unit has the characteristics of an open unit, the analyst can rather safely hypothesize that it will engage in fairly cautious, deliberative behavior that will not commit many of the resources of the government and be more cooperative in tone. In effect, the closed unit bears within it the seeds of innovation, change, and unilateral initiatives as well deadlock. The open unit reflects more of an interest in maintaining the status quo, being relatively incremental and provisional in its behavior.

## HOW EASY IS THE FRAMEWORK TO USE?

A major value of the case studies has been to aid in the identification of problems, both theoretical and mechanical, in the framework. Both the specialists and students who prepared the case studies took seriously the request to consider how easy it was to apply the framework to a historical situation. While none found it difficult to use the framework, they did raise a number of questions as they worked through the case materials. Their observations fall into four categories: (1) the information required by the framework, (2) the focus on a single occasion for decision, (3) the specification of the boundaries between types of decision units, and (4) the relevance of theory to defining what happened.

### *Information Required by Framework*

A reading of the pieces in this special issue that describe the various types of decision units suggests that in applying the framework the analyst needs rather detailed information about the decision-making process. And, indeed, it is for this reason that many scholars of foreign policy and international relations have precluded studying how decisions are made. They argue the information is too hard to find and often, when available, is incomplete. Thus, we were fortunate to have among those doing case studies for us individuals who had studied a particular case extensively and in that examination paid close attention to the policymaking process. These experts pushed us to clarify the criteria for making certain judgments and facilitated the elaboration of the guides now provided to analysts starting to apply the framework. But they also complemented the framework for its focus on the use of key contingencies to differentiate among types of decision units. By distinguishing how sensitive a predominant leader is, where the primary identity of members of a single group lies, and if a coalition has rules or not, the framework points the analyst in a particular direction and specifies the kinds of information that will be needed to indicate what is likely to happen. Once these initial distinctions are made, only certain kinds of information are necessary to denote the process and substantive outcomes. An attempt has been made in the framework to limit the analyst's task after the initial specification of the decision unit. In effect, the decision unit framework helps the analyst denote whose positions are probably going to count in the policymaking process for a particular problem, thus narrowing down the number of people and groups that have to be studied.

An examination of the graduate students' cases that were completed in the past several years indicates that the ease in collecting detail about the policymaking process has increased as more countries and governments use the Web to provide current information about events and policies. Although some care must be taken to ensure the authenticity of the information, students have been surprised by what is available on how decisions were made. Moreover, the media has become more involved in analyzing particular decisions, interviewing those who participated in the decision-making process and overviewing available documents. Autobiographical and biographical materials are increasing as policymakers' points of view and involvement are documented. Because, however, it still remains easier to report on and study the decisions of one leader than to examine what goes on within a group or coalition, more of the materials that are extant provide information about predominant leader decision units than those involving single groups or coalitions of actors.

In applying the framework to the sixty-five cases, we have been reconstructing history. That is, those engaged in the case studies have used the framework to examine decisions that have already been made and about which we have some records. Another way to think about the framework is to envision it as a forecasting tool, as a means of understanding what a government is likely to do in response to a foreign policy problem. A number of the experts and students in the present study observed that once having applied the framework they recognized that it provided them with a tool for organizing new case study information. Indeed, one (Sundelius, Stern, and Bynander, 1997; see also Stern, 1999) has found parts of the framework relevant to examining crisis management in a number of European countries. Could it also serve as a way of collecting material regarding a current case? Knowing something about the structure of the government facing the problem, could we ascertain what decision unit is likely to deal with such issues and begin searching for information to answer the appropriate questions regarding that particular type of unit? Where information is sparse, we could follow several paths in ascertaining the nature of the process outcome to see which best matches what actually appears to happen. When a governmental decision on the problem became evident, we could compare our prediction with the decision that was made and examine what information did, or would have helped, in making the correct inferences.

## *Focus on a Single Occasion for Decision*

The decision units framework is triggered by a foreign policy problem facing a government and, in particular, a specific occasion for decision that requires action. Those writing the sixty-five case studies examined here had little difficulty in isolating occasions for decision. In fact, most cases involved a number of such occasions; in several it was possible to identify over twenty occasions for decision. The data indicate that policymakers make multiple decisions in responding to a problem. In reality, what scholars often identify as a single case is composed of a variety of points in time when some decision unit is faced with making a choice. And often the kind of decision unit responsible for taking action changes across these occasions for decision in the course of dealing with a problem. Consider the number of occasions for decision that faced the new Bush administration in the spring of 2001 as the U.S. government tried to get its crew and plane back from Chinese soil after a collision with a Chinese fighter off the South China coast. As observers of this event, we counted fifteen.

Although we instructed those doing the cases to explicate only the occasions for decision that resulted in authoritative actions on the part of the government, it became quickly evident that such decisions are not made at a single moment in time when policymakers meet and resolve the issue. Instead, there can be multiple such occasions for decision that cross days, weeks, and even months—if not, at times, years. By applying the framework to these different occasions for decision, the analyst gains a perspective on the twists and turns that policymakers engage in as they try to cope with the problem facing them. Analysts also learn what decision units were involved in working on the problem and how the type of unit may have changed across time. The multiple occasions for decision described in the case studies being discussed here suggest the viability of injecting a dynamic element into the decision units framework by focusing on the sequence of choices that appear to be contained in many foreign policy problems. Understanding how the results from one occasion for decision influence and/or mingle with information from the environment to shape the nature of the next occasion for decision can provide us with insights about the choice-making process across time. Whereas examination of a single occasion for decision provides us information about policymaking within a particular type of decision unit on an aspect of the problem, examination of multiple occasions for decision enables us to learn about the government's strategy and way of coping with the more general problem. Like the meteorologist, studying one occasion for decision is similar to giving the weather forecast for tomorrow in a specific city, while exploring multiple occasions for decision suggests what to expect the general pattern of the weather to be over the next week or month. At issue is how does what happens in each occasion for decision influence what goes on in the others. We will return to this topic in a later section when we consider how to make the framework more dynamic.

## *Specification of Boundaries Between Types of Decision Units*

Although a lot of effort went into distinguishing the conceptual boundaries between the three types of decision units and the conditions of the key contingency variables, the case study authors raised some further concerns. Their problems grew out of the framework's focus on involving the informal decision structures in foreign policymaking not simply the structural arrangements formally stipulated by a constitution or the like. Some examples will illustrate the issues.

What if we have a fairly powerful leader and his advisers meeting in response to an occasion for decision. And assume for the moment that the leader is sensitive to contextual information and, thus, interested in input from the advisers. Is the authoritative decision unit in this instance a sensitive predominant leader or a single group with a dominant leader whose members are loyal? The critical issue here is the role of the advisers. Are they simply providing advice and information which the leader will take and use in making his or her own decision or do the advisers wield authority within the group such that they constrain what the leader can do—that is, can they say no to what the leader wants and have their decision become governmental policy? Sometimes scholars can disagree over the answer to this query. Take, for instance, Britain's decisions during the Munich crisis. While some (e.g., Walker and Watson, 1989; Walker, 1990) argue that Chamberlain operated as the leader of a single group (the inner cabinet) whose members' primary identity was to the group in this crisis situation, others (e.g., Colvin, 1971; Middlemas, 1972) conclude that Chamberlain was a predominant leader who used the inner cabinet to legitimate his own decisions. For this case, analyzing the five occasions for decision that occurred during the height of the crisis reveals "a pattern of Cabinet decisions rather than Prime Minister's Rule" (Walker, 1990:23). Indeed, in the fourth and fifth occasions for decision the cabinet overruled Chamberlain's preferences.

Another problem in distinguishing among decision units revolved around the presence of "outside experts" (e.g., the military, financial consultants, regional specialists). What is the role of experts who are invited into a group's deliberations and appear to have a significant impact on what happens but are not generally a part of the group? The answer to this question has implications for the single group decision unit when unanimity is required. And it can suggest that what usually is a single group decision unit may become a coalition of autonomous actors for a particular occasion for decision. Consider the roles that the ministers responsible for Cyprus and EU affairs played in the Turkish inner cabinet's decision in 1999 regarding whether or not to accept candidacy in the European Union that carried with it certain stipulations not applied to other candidate states relating to Cyprus and its Aegean disputes with Greece (Cuhadar, 2000). Both ministers participated in the discussions concerning what the Turkish government's response should be even though they were not official members of the decision-making group. Unanimity was the decision rule governing this particular group composed of the leaders of the three parties who had formed the government and the minister of foreign affairs. An examination of the decision-making process indicated that unanimity was required only among those who usually made up the group because as leaders of their parties each had the capability of bringing the government down if he or she did not agree with the decision. Moreover, for the most part, this inner circle had been the group that had handled the negotiations with the EU over candidacy. The "extras" in the meetings were viewed as experts on the issues under discussion; they served in an advisory capacity to provide information the others needed in pursuing a decision. This case suggests that an important criterion in determining the role outside experts are playing revolves around whether or not they can block or reverse the decision. Even though the minister for Cyprus disagreed with the decision the inner cabinet made, he did not have the authority or legitimacy to change it once made nor to be considered among those whose acceptance was necessary for there to be unanimity.

A third cloudy area in distinguishing among types of decision units comes when we are faced with a coalition cabinet. Because coalition cabinets in parliamentary systems are generally composed of representatives from different political parties, they are usually considered coalitions of autonomous actors. Yet, at times, they can be viewed as single groups. Often such coalition cabinets become single groups when the government faces a threat to its survival or legitimacy—for instance, there is a suicide bombing or Arab attack on the Israeli government which has in recent times been led by a coalition; there is a challenge to the neutrality principle of Sweden or a foreign submarine trapped in its waters. Quick decisive action is demanded and differences among parties and groups become more muted. At issue here is how necessary is it that the cabinet members check back with their party leadership before presenting or considering a proposal before the cabinet. If the cabinet member cannot make a decision on his or her own without checking first with the party leadership, the decision unit is a coalition. If, however, the cabinet member can act more autonomously—he or she is the leader of a party—we argue that the decision unit is a single group.

## *Relevance of Theory to Defining What Happened*

The variables that become relevant to examine for each of the decision units in the decision-making process were derived from bodies of theory that have emerged from several decades of research examining the impact of leaders, groups, and coalitions on governments' foreign policy. These theories have been detailed in the articles in this special issue on each of the types of decision units. The key contingencies were selected because they differentiated among the various theories that have developed around these three types of units. We were interested in whether the case studies reinforced the theoretical insights already built into the variables in the framework, provided us with new variables that elaborated previous theory, or indicated lack of support for a theory. Using these questions as a guide, we have examined the "decision paths" designated in the previous articles for the fifty-four cases where there was a fit between the process outcome from an application of the framework and the historical outcome. Generally there were three to four cases to study for each of the types of decision units after the key contingency variables were applied.

***Predominant leader.*** For the most part the twenty-two cases where there was a match between framework-determined and historical outcomes for predominant leader decision units also reinforced theory. The best fit occurred for the highly insensitive leaders—the crusaders—who exhibited expansionistic or evangelistic leadership styles. Figures like Saddam Hussein and Slobodan Milosevic followed expectations in the cases that were examined. The worst fit between theory and outcome was for the moderately insensitive leaders—the strategists—who though they challenge constraints do so by testing the political winds and seeking out a range of information. These leaders have goals but the means of achieving them are determined by the nature of the political context and what seems feasible at that point in time. Hafez al-Assad, F. W. deKlerk, and Anwar Sadat are examples of such leaders who were studied in the cases. It is easy for the analyst to expect a more extreme action than the rather moderate, low-risk responses that can result. Important in considering what strategic leaders are likely to do appears to be knowledge in the particular situation about what are relatively "safe" bets politically that will facilitate movement toward a goal. If there are none, inaction is a real possibility. Moreover, strategic leaders are highly influenced by the people who provide them information about the current political context; thus, examining these individuals' views can aid in determining what may be viewed by the leader as safe politically.

***Single group.*** In 80 percent of the twenty groups examined in the case studies that matched the historical outcomes, at least half the variables identified as relevant to understanding the decision process in single-group decision units were included in the analysis. And in most instances the case studies that determined a single group was the authoritative decision unit and were predicted accurately by the framework fit theoretical expectations. Those that differed from the theory reinforced our view that there are alternative processes that allow groups to bypass the so-called pathological outcomes of group decision making proposed in discussions of groupthink and bureaucratic politics.

In several cases, the single-group decision units whose members' primary identity was to the group were able to break out of the groupthink mold because group norms permitted serious examination of other options or there was an individual in the group who forced the leader to review other alternatives. Consider the role Clark Clifford played in President Johnson's Tuesday Lunch group during the Vietnam War. It was difficult for Johnson to continue disregarding Clifford's growing distaste for the war given his general admiration and respect for Clifford's ideas in general. Here was somebody the group thought highly of who was not going along with the majority and calling the options the group was discussing into question. If the group was to continue to function, members had to listen to Clifford and consider how to incorporate his views into the group's decisions.

Single groups that must reach unanimity before action is possible and whose identity lies outside the group appear able to escape the so-called resultant effect concomitant with bureaucratic politics through the presence of a broker. These are members who have no personal or organizational stake in what happens with regard to the issue under discussion but have the ability to question others closely about the relative merits of their options and to push for evidence for their interpretations of what needs doing. The Soviet Politburo's decision to resume sending offensive weapons to Egypt in January 1973 was the result, in part, of the efforts of Suslov to act as a broker on this issue in return for a quid pro quo in the future (Stewart, Hermann, and Hermann, 1989). The Brezhnev Politburo had the norm of building consensus/unanimity before taking action.

***Coalition.*** In 75 percent of the twelve coalitions examined in the cases studies that matched historical outcomes, over half the variables identified as relevant to understanding the decision process in coalition decision units were included in the analysis. And in most instances the case studies that determined a coalition was the authoritative decision unit and were predicted accurately by the framework fit theoretical expectations. The couple that differed from the theory reinforced our view that pivotal actors, particularly those that control relevant resources, have an important role to play in the decisions that coalitions make. In coalition decision units there are two types of pivotal actors. In considerations of coalitions where the majority can rule, a pivotal actor is a member for whom "the absolute difference between the combined votes (weights) of members on his right and of members on his left is not greater than his own weight" (De Swaan, 1973:89). Any policy agreement must necessarily include this actor. Or, in coalitions where there are no rules and anarchy prevails, the pivotal actor is that person, group, or institution who controls disproportionate amounts of key political resources. Such actors can shape the nature of the process by pushing for their preferences or by acting as a mediator.

An illustration of the role a pivotal actor can play is found in examining the Turkish government's reaction to the overthrow of Makarios on Cyprus and a threatened Greek annexation of the island in the summer of 1974 (Ozkececi, 2000). The coalition involved in responding to this occasion for decision included the cabinet, the Joint Chiefs of Staff of the military, and the National Security Council; the coalition could take action with a majority supporting a response. At the outset, however, there were hot debates within all circles as to the appropriate action. With authority in or over each of the members of the coalition, Prime Minister Bulent Ecevit attended all the meetings of the various entities, keeping all informed of the others' positions and arguments. Using this personal diplomacy, Ecevit was able to build a majority among the members of the coalition for intervention into Cyprus. In this instance, the prime minister used the authority he had to interact with all members of the coalition to press a position that he favored and to gradually counter the arguments against his preference.

## How Comprehensive Is the Framework?

Of interest in the development of any framework are concerns over capturing the essence of the domain one is trying to model. A valuable outgrowth of the collection of case studies that we now have applying the framework is that their authors have considered if there are major elements missing from the framework—if we have inadvertently excluded important decision-making dynamics from consideration. They have made some highly relevant suggestions.

But before turning to these proposals, let us reiterate that the decision units framework is not intended to be a general explanation of foreign policy. The focus, instead, is on foreign policy decision making at the point of choice. The framework examines the policymaking processes of those actors (powerful leaders, single groups, or coalitions of actors) who have the authority to commit the resources of the government with respect to a particular occasion for decision without having their decision reversed. International and domestic factors are only included as they are interpreted by, or represented in, the members of the decision unit. Moreover, the framework currently does not explore how the problem gets defined or how decisions get implemented. In essence, the framework is intended to aid in understanding how foreign policy decisions are made by those with the authority to commit the government to a particular action or set of actions when faced with an occasion for decision.

Having thus delimited the focus of the framework, the case study authors argue that we inadvertently let in the "back door" some of what was supposed to be excluded. There are the "outside experts" who are not part of the official, or even unofficial, authoritative decision unit with power to make a decision without reversal and yet, nonetheless, help to frame the problem and the options being considered by those with authority. Then there is domestic political opposition that, if strong, can shape the expectations of what is possible in the authoritative decision makers' minds if the latter hope to retain their positions and legitimacy. And what about the degree of uncertainty the authoritative decision makers have regarding the nature of the problem that is facing them; when is a problem sufficiently defined to take action? Moreover, how do cultural and political norms restrict the kinds of choices policymakers can make regardless of the nature of the decision unit? Let us explore these concerns in more detail.

A number of the issues raised by the case study authors center around the definition of the problem phase of the decision-making process and, in particular, who or what gets to frame the way the problem is viewed. In certain instances, those who discover something has happened have the initial chance to define the problem, however accurate their representation is. Thus, a Swedish nuclear plant manager perceived that his plant had had an accident because of the increased radiation on his employees when, indeed, the radiation was coming from the Chernobyl plant in the Soviet Union (Stern, 1999). It may take time to re-frame the problem once policymakers lock onto an initial perception of what is occurring. Who the authoritative decision makers are may, in turn, shift as the problem definition changes. Similarly, domestic political opposition may inadvertently help to transform how a problem is represented. Consider the effect of the opposition voices that crowded into a stadium to listen to Clinton's foreign policy team discuss the relevance of military air strikes against Iraq in the winter of 1998 weapons inspection crisis (Swords, 2000). In what the Clinton White House had considered a safe midwestern venue, their foreign policy team was roundly criticized and on national television. What was supposed to be merely reporting a decision had the impact of re-shaping the nature of the problem and what options were viable.

Certain cultural and political norms can limit the kinds of options that can be considered in dealing with a problem. Indeed, problems and options may be linked. For example, the Israeli image of certain Arab states and groups as enemies translates into conflictual behavior when policymakers perceive even the slightest threat. It takes little activity on the part of these Arabs to be viewed as threatening. The norm of Swedish neutrality is another case in point. Under such a norm, violent options are almost unthinkable in response to a foreign policy problem unless one's survival is at stake; only those alternatives that maintain neutrality are feasible.

Just like cultural and political norms, the expectations and experiences of the policymakers can also influence not only how problems are framed but the types of alternative actions that can be pursued. In the presentation of the material on the coalition decision unit, Everts discussed the cross-pressures facing the Dutch cabinet in the highly controversial cruise missile case. The concerns that Dutch policymakers had regarding their own futures probably were more important in defining the options that the cabinet viewed as available to them than the external pressures from their NATO allies. Although the parliamentary opponents and leaders of the antinuclear movement did not participate in the proceedings of the decision unit (nor were they really in a position to remove the cabinet from power), by their actions they were able to shape the policymakers' expectations of what was possible. So-called outside experts invited to provide information and options to members of the decision unit can also shape what the decision makers view as feasible. And depending on which experts are included or excluded from participation, certain options can be highlighted or not discussed at all (see Hoyt and Garrison, 1997).

What we have been describing here seems to be an important characteristic of most decision units when faced with a problem: they engage in reducing their uncertainty regarding the nature of the problem and the options they have for dealing with the problem. And they use a number of different tactics to delimit what they are experiencing. Moreover, policymakers may buy time to search for more information through fairly minimal or maintenance-oriented foreign policy behavior when the occasion for decision is perceived to involve a high degree of uncertainty. As a result, a decision unit may be involved in more occasions for decision when the problem is uncertain and combine the definition of the problem and choice-making stages of the decision process together or work back and forth between the two. Several of our case study authors have proposed that the decision units framework may be more appropriate for those occasions for decision when the members of the decision unit believe they understand both the nature of the problem and their options and are ready to act.

These challenges to the decision units framework all center around the fact that a decision unit does not come to each occasion for decision as a tabula rasa acting as if it were, in essence, a brand new problem. An important next step in making the framework more comprehensive in detailing the decision-making process involves constructing a more dynamic model by examining a sequence of occasions for decision. In the process, we can begin to elaborate the parts of the framework that now appear truncated.

## TOWARD CONSTRUCTION OF A DYNAMIC FRAMEWORK[4]

The decision units framework was designed around the occasion for decision as the basic unit of analysis because it denoted that period of time when the authoritative policymakers responsible for handling a specific foreign policy problem remained constant. The occasion for decision is like the single frame in a movie. When new information is perceived that calls for another round of decisions, we have a new occasion for decision and, perhaps, another type of decision unit. And just as with a movie, we recognize as an outgrowth of the sixty-five case studies that it is critical now to devise a means of putting the film into

---

[4] This section builds on two papers by Hermann and Billings (1993, 1995). The authors would like to thank Robert Billings for his many insights regarding sequential decision making.

motion and explaining how we get from one frame to another. We need to locate discrete occasions for decision into a dynamic process that captures a particular decision unit's experiences with prior choices and relevant intervening developments. Should the authoritative decision unit shift in an ongoing problem, it is important that we can identify to what degree, if any, these policymakers are likely to build upon, or react to, the actions of the previous decision unit. In essence, we are interested in the relationship between the decision units framework and sequential decision making.

Thoughtful scholars have long acknowledged that policymaking is itself an ongoing process in which decisions (and nondecisions) generate responses that create new opportunities for choice (e.g., Lindblom, 1959; Snyder, Bruck, and Sapin, 1962; Brecher, 1972; Steinbruner, 1974; Axelrod, 1984). Indeed, recognition of the serial nature of decision making appears in the literatures on cybernetics, incrementalism, game theory, escalation of commitment, and risk taking. But generally after raising the issue most studies include only one or two iterations or simply treat the actual sequence as a given (see Stevenson, Busemeyer, and Naylor, 1990). As a result, the following discussion is necessarily preliminary and awaits further exploration, as does the notion of sequential decision making in the study of foreign policymaking more generally.

In the following discussion we will consider that sequential decision making occurs when policymakers engage in a series of decisions about the same problem across a period of time. Decisions (to take certain actions or to do nothing at present) are made in response to an initial occasion for decision. With these first steps comes a momentary sense of closure and those involved turn their attention to other matters. The decision making becomes sequential when these same actors find themselves faced with reconsidering the problem or some variant of it. In elaborating the decision units framework, we seek to understand the circumstances under which prior decisions are allowed to stand, are adjusted, or are completely altered. To help with this task, we need the answers to two questions: (1) When will a decision unit reconsider a prior decision about an ongoing problem rather than continue the status quo without deliberation? (2) When a decision unit does reconsider a prior decision, under what conditions will it change its earlier position?

## *Reconsideration vs. Continuation*

In some instances policymakers may recognize that their current decisions will necessitate future decisions on the same problem. When the Israeli cabinet decided to send Foreign Minister Eban to consult with the American government in the face of the rapidly deploying Arab forces just before the Six-Day War in 1967, they knew there would be further decisions to make in the near future. In other circumstances, those in the authoritative decision unit may expect that they have acted definitively on a problem. Probably the Dutch cabinet believed in November 1985 when they finally were able to agree on the deployment of the new cruise missiles on Dutch soil that they had at long last disposed of what had seemed a never-ending problem. So, in these latter cases, what leads the decision unit to take up the issue yet again?

Moreover, reconsideration of a previous decision or set of prior decisions is only the first step. When policymakers do reconsider a decision, will they change their prior position? A decision unit may reconvene to review its past actions and subsequent developments, concluding that no new decisions are appropriate. Or they may elect to redouble their efforts and increase their commitment to the course previously established. Alternatively, they might choose to change direction, that is, to terminate the prior action and follow another course. If a new action is taken, it may vary from past actions in both magnitude and direction. It might involve only a minor adjustment or clarification in the previous activity or it could entail a major increase in the commitment to the prior act. And the decision unit could initiate a complete break with the previous behavior and reverse their policy direction. During the long course of the American involvement in Vietnam, all these different types of responses occurred at one time or another.

From an analytic point of view, we can suggest a number of circumstances under which policymakers logically could be expected to reexamine a prior decision and modify the actions they took earlier. If the prior action was *successful* in achieving its objectives, then the decision unit might reconvene to reallocate resources to other tasks. Conversely, if the implemented action appeared to be a *failure*, the decision unit might wish to determine whether another course could better achieve the intended results or should they consider foregoing the objective. Sometimes actions that appear to be effective with respect to a particular purpose also have serious *unintended consequences;* policymakers may decide to review their earlier decisions because the costs of implementing the action have become too high or its side effects too costly. In other words, feedback from the environment can lead to reconsideration and possible change in the prior response to the problem. The feedback can cause the decision unit to alter its definition of the problem and, as a result, to make different assumptions than those they followed in the previous decision(s).

An examination of the sixty-five cases in the present study indicates that responding to feedback from the environment is more characteristic of the so-called open than closed decision units. The units that depend on information from the context in making their decisions—the more sensitive predominant leaders, the single groups whose members' primary identity resides in outside entities and have the need to reach consensus, and the coalitions convened under a rule of unanimity—are more likely to use the responses to their actions as a guide to next steps. The closed units—the other eight types of decision units—are going to be more selective in their perceptions of the effectiveness of their policies and actions. These units are more likely to resist reconsideration of their decisions and change. Unless such decision units themselves change as a result of a reshuffling of who is involved or there is a regime change, the emphasis will be on stability and continuation of policy rather than change. Only dramatic and unrelenting failure or quite costly unintended consequences are likely to jolt such units from the status quo.

The case data suggest that rules governing sequential decision making may differ depending on the nature of the decision unit. The question becomes under what conditions might the open decision units become more resistant to feedback from the environment around them and the closed decision units become more willing to deal with this same feedback? The literature on decision making in politics and organizations suggests several factors could play a role in reversing the ways these different types of decision units respond to feedback: expectations, agitators for change, and accumulated experience.

## *Expectations*

In their selections of a response to an occasion for decision, decision units develop expectations about what the action will accomplish. They may believe that it will yield more information; reveal an adversary's intentions; generate support for their position; or slow, stop, or reverse the effects of the problem. Policymakers' expectations indicate when they are likely to recognize and interpret new information from the environment that can trigger further consideration of the problem. An analysis of our sixty-five case studies suggests that the actions range along a continuum reflecting the decision units' relative certainty on how to interpret and cope with the problem they are facing. At one end are actions that involve further search for information and at the other are definitive actions that are intended to resolve the problem. In between are provisional actions that are tentative and limited in their scope and conditional actions that demand a response from some other indicating whether future decisions are necessary.

When decision units initiate a search for information, they expect feedback. Indeed, members are interested in improving their understanding of some aspect of the problem. In some instances the policymakers may have formed a hunch about the problem they seek to confirm through the search. Regardless, they are proactive in canvassing the environment for data that can help them deal with their uncertainty. Unless the search is a smoke screen for inaction, the decision unit knows that it will be deliberating further on what to do about the problem when an answer is forthcoming. Even if the search is unsuccessful, there will be a need to reconvene. Anticipation of future reconsideration of the problem is built into the search decision.

When decision units choose provisional actions, they perceive a need to make some effort to address the problem but in a tentative and limited fashion. Since the policymakers remain uncertain about the effectiveness of the response that was chosen, provisional actions are generally small incremental steps that keep options open, including the ability to terminate the activity or limit its consequences if it does not seem to be working. Decision units that implement provisional actions are sensitive to information from the environment that indicates how successful their actions have been in resolving the problem. They are prepared to reconvene if the steps they have taken do not appear to be having an effect.

In undertaking conditional actions, decision units expect a response from the environment. Such actions signal that future behavior will be contingent on what others associated with the problem do. By initiating conditional action, the decision unit has probably accepted a certain causal interpretation of what is happening that enables them to specify what certain others must do (or stop doing) to alleviate the problem. Positive or negative incentives are offered for compliance with the action. The expectation is that there will be a response establishing whether the appropriate others have conformed to the directives. If the response is as anticipated, there may be little need to reconsider the previous decision. If the conditional action is unsuccessful, the decision unit is likely to reconvene.

Decision units that use definitive actions believe they have adequately diagnosed the problem and have an effective means of dealing with it. Such policymakers see no reason to make their activity dependent on others or to defer what appears to be a reasonable way of resolving the problem for more information. Confidence in their own ability to cope with the problem is likely to reduce this decision unit's sensitivity to negative feedback and provide them with little incentive to reconsider the issue. So far as the decision unit is concerned, the matter is settled.

As this discussion implies, members of decision units appear to become more certain about the appropriateness of their reactions as they move from search to definitive actions and the actions themselves become less cautious. Decision units are more committed to their current course of action as the behavior changes from that of searching for more information to being provisional to acting conditionally to being definitive. With increased certainty and commitment goes reduced alertness to signals that things might not be going well and greater reluctance to changing course if forced to reconsider. Indeed, research (e.g., Staw and Ross, 1987; Brockner, 1992; Uhler, 1993) has found that decision units with a strong commitment to a previous action tend to increase their commitment to that action when confronted with negative feedback rather than change course. They engage in an escalation of commitment by investing more effort in the goal of the prior action. Moreover, policymakers are more likely to pay attention to information from the political environment the more it conforms to their prior expectations. In fact, these expectations will dictate the kinds of information they will regard as relevant. Thus, whereas decision units engaged in a search will be interested in answers to their questions, those involved in provisional and conditional actions will be looking for the impact of the action on others and these others' response. Furthermore, the more strongly committed a decision unit is to its prior action, the more likely members are to interpret information from the context as supportive and not to recognize disconfirmatory information (Vertzberger, 1990). Taking a definitive action may diminish a decision unit's vigilance for feedback.

The case data in this study indicate that the initial actions of a greater percentage (61 percent) of the open types of decision units were of the search and provisional kind while the larger percentage (64 percent) of the initial actions of the closed types of decision units involved conditional and definitive behavior. The open types of decision units do, in fact, appear to be more interested in feedback from the context than do the closed types. And there is some indication in those cases where the analysts examined more than one occasion for decision that the open units were involved in more occasions for decision as they worked to deal with a problem than the closed units. Across all the decision units, predominant leaders who were crusaders and single groups whose members' primary identity was to the group engaged in the most definitive actions. Based on the theory behind these two types of decision units, they are likely to be the most confident and committed to their actions and ready to move on. But they are probably also the most willing to engage in escalation of commitment if things appear to be going wrong. Search and provisional activities were most frequent among the predominant leaders who were opportunists and the coalitions without defined rules of the game. In the case of the opportunists, they need constant information on which to base any decision; in the case of the coalitions without rules, members are often so focused on deciding who is in charge that they can only act provisionally. The foreign policy problem may be used in the struggle for power.

Among all fourteen different types of decision units, predominant leaders with a strategic leadership style and single groups as well as coalitions that allowed for majority rule evidenced the most variability in the kinds of actions they pursued and the number of occasions for decision in which they participated in the course of dealing with a problem. The strategic predominant leaders appear to have used search and provisional activities as "trial balloons" to see what response they might encounter should they choose a particular action. Conditional and definitive actions only followed when these leaders were confident they were in control of what was likely to happen. Consider Egyptian President Sadat's circuitous route to the October War in 1973 detailed in the discussion of the predominant leader decision unit. Because who is in the majority can change in single groups and coalitions with this decision rule—and, indeed, those who once were included in the decision unit can be excluded because of their positions on a particular issue—it is probably not a surprise that such units evidenced a variety of kinds of action. The presence of some who lose out means that there are always members interested in monitoring what is happening and pushing for reconsideration; there are always potential agitators for change present in the group or coalition.

## *Agitators for Change*

In describing the process outcomes that result from the application of the decision units model, we noted the asymmetric quality to some of the outcomes and the fact that such decisions are rather unstable. Built into certain process outcomes is the notion that members of the decision unit are likely to want to revisit the decision later on. When one party's position prevails, there is a lopsided compromise, and participants in the decision unit are engaging in fragmented symbolic activity, some participants in the decision unit do not "own" the decision that is made and have a reason to monitor resulting action and agitate for reconsideration of the decision should feedback be negative or not have the effect they want. These individuals are motivated to monitor the environment for signals relevant to a given problem because of their discontent with the prior action. In the process, these agitators can become advocates for change.

As was observed earlier in this special issue in the description of the decision units framework, these process outcomes are thought to be more dominant for certain types of decision units. Having one party's position prevail is proposed to be more characteristic of the relatively insensitive predominant leaders, single groups with outside loyalty and a majority decision rule, and coalitions with a norm favoring majority decisions. Lopsided compromises were expected to occur more often with moderately sensitive predominant leaders when their preferred option was feasible. Fragmented symbolic action was considered more likely in coalitions with no established rules and high political instability. And, in fact, an examination of the sixty-five cases in this study lends support to these assignments. A little over half of the outcomes for these decision units were of the predicted type. Thus, agitators for change may cluster in particular kinds of decision units where they believe themselves to have had less chance to influence the outcome than others in the unit. And like a George Ball or Clark Clifford in U.S. decision making on Vietnam, the current hard-line members of the inner circle of the Chinese Communist Party with regard to partnership with the United States, or the reformers in Iran on economic opening to the West, they may push for reconsideration of decisions that go against them when events support their original positions.

These agitators for change often represent a minority position among advisers, in a group, or in a coalition. By minority we mean individuals with divergent perspectives and preferences for action as well as competing definitions of the problem and occasion for decision from those of a majority. Research on minority influence over the past two decades in social psychology and foreign policymaking has shown that there are certain conditions when the agitators being discussed here have the possibility of bringing about the desired change (see Kaarbo, 1998; Kaarbo and Beasley, 1998; and Kaarbo and Gruenfeld, 1998, for reviews of this literature). Minorities appear to have an effect when they argue their position consistently over time (Mugny, 1975; Moscovici, 1985), are moderate in size (Wood et al., 1994), appear to focus on the group's interests rather than self-interest (Eagly, Wood, and Chaiken, 1978), and have support in the larger societal context (Clark, 1988; Mugny and Perez, 1991). The research suggests that the impact of minority views takes time to develop. But after hearing the minority persist in repeating its views and challenging the majority's ideas, members of the majority may begin over time to reappraise the situation and develop new perspectives—while different from their prior view, not necessarily those advocated by the minority. Such reappraisal is particularly likely when the group receives negative feedback on its previous actions.

This literature on minority influence has implications for the effects that agitators for change may have on future actions of decision units. Because such members have reservations about the merit of the prior decision, they are more likely than those whose positions are represented in the choice to recognize discrepant feedback from the environment, particularly if such feedback does not fulfill the earlier expressed expectations of members of the decision unit. By assuming a postdecision monitoring function, these members increase the likelihood that the majority will be forced to acknowledge negative feedback and reconsider the problem. And if said agitators persist in their message, retain a consistent position, and argue their position is for the good of the decision unit given what is happening to the prior behavior, they may persuade enough members to reconstitute the majority. Consider the change in Soviet policy toward Czechoslovakia in 1968 when the interventionists on the Politburo who were the minority in July maneuvered to become the majority by agitating for their position in a persistent manner and by presenting information that demonstrated a deteriorating situation in Prague (Valenta, 1979). Alternatively, the agitators may induce some members of the majority to engage in divergent thinking and create a new position that can gain support among both the majority and the minority (Wood et al., 1994). In either case, the agitators for change have pushed the majority to reconsider its position.

Even if the agitators for change are not successful in influencing the majority to modify their prior position, they may have some effect indirectly on the majority by reshaping the latter's values or decisions on related issues. Halperin (1972, 1974) argues that such indirect change occurred in U.S. decision making surrounding the deployment of the antiballistic missile (ABM) system in the late 1960s. Secretary of Defense Robert McNamara took a position in opposition to the majority of the bureaucrats who favored rapid deployment of the ABM. McNamara was not successful in delaying deployment but did have an effect on long-term strategic thinking. Indeed, Halperin attributes U.S. acceptance of ideas regarding nuclear sufficiency rather than superiority and the talks with the Soviet Union on limiting ABMs to McNamara's earlier efforts.

Agitators for change may not be limited to members of the decision unit. Decision units are ultimately accountable to certain constituencies be they voters, special interests, other leaders, the bureaucracy, or a legislature. They have to justify their decisions to these others. A number of factors can affect the relationship between a decision unit and relevant constituencies: the ability of the constituents to assess the performance of the decision unit, their access to the decision unit in order to convey their preferences, the nature of the feedback that is perceived by the constituents about the decision unit's performance, and how accurately such feedback is interpreted. Knowing that an election is near, that one's approval ratings are low, that the military needs very little excuse to stage a coup, or that the media is investigating will probably enhance the effects that outside constituencies can have on limiting the options the decision unit can consider at that point in time and may even truncate work on a particular problem. Certainly if policymakers discover a major discrepancy between their past actions and the preferences of powerful constituents, major changes are likely. But in order not to appear inconsistent, the changes will probably be by expansion or elaboration rather than explicit reversal.

At issue is what happens when the decision unit's actions are ineffective or failing and these constituents start to lobby for change, for example, when the U.S. marines died as a result of the bomb explosion in their barracks in Lebanon or the U.S. soldiers were killed and their bodies dragged through the streets in Somalia. The decision unit is forced to reconsider its actions and to either find some external cause on which to affix blame or accept responsibility, while at the same time outlining steps they are about to take to deal with the difficulty. In assuming accountability, however, the decision unit is likely to assign the failure to inadequate implementation (e.g., choosing the wrong solution or not using enough resources) or to unsupportive minorities in the unit—attributions that do not challenge the diagnosis of the problem or the unit's goals and can be adjusted (see Wong and Weiner, 1981; Salancik and Meindl, 1984). Here is where we may see a difference among types of decision units. Believing themselves more accountable in general, the open units may be more willing to attribute failure to a lack of goal clarity or incorrect problem diagnosis than the closed units and, as a consequence, to engage in more extensive reconsideration of their prior positions and a potentially more dramatic shift in direction.

The closed units will probably stick to their original interpretations while slightly modifying what they were doing. Agitators for change within the decision unit may increase in stature within the open units but may be excluded, isolated, or used as a scapegoat in closed units. "I told you so" is not what the closed units want to hear.

## *Accumulated Experience with the Problem*

The history that a particular decision unit has with a problem can also influence when actions will be reconsidered. If a problem is being confronted for more or less the first time, a decision unit is likely to be tentative and their actions will probably involve further search for information or be provisional. New information (even if negative) will be eagerly sought and used to shape policymakers' emerging views. Beliefs about the new problem are less likely to be well-established, nor are governmental organizations likely to have entrenched positions and procedures for dealing with the problem. As the problem is encountered repeatedly, decision makers are likely to develop views about what is happening, why, and what needs to be done. Of course, the nature of some problems may be so perplexing and illusive that policymakers' convictions about preferred modes of treating them may never develop or are held with low confidence. For several U.S. administrations, the Israeli-Palestine conflict seems to have had this quality. But for many problems, the decision unit develops a set of routines for dealing with them and resistance to further review and examination begins to increase. Beliefs about the nature of the problem become well established, investments are made in particular actions, and a number of disincentives—programmatic comparisons of costs and benefits, personal esteem, institutional practices—kick in to encourage the unit to stay the course and conclude that little new can be gained by another round of deliberation.

As these convictions about the nature of the problem and the means of coping with it become stronger with successive decisions, it will probably take the occurrence of a very costly and highly visible failure or a major success with recognizable reorganizing consequences to get the decision unit to reconsider what it is doing and undertake change. If negative feedback persists over an extended period of time during which no further reconsiderations have occurred, we postulate that the resistance to re-addressing the problem may decline. Such is particularly likely if the negative feedback is becoming newsworthy, if such feedback is evident to important constituencies, or if the set of agitators for change within the decision unit has persistently opposed the present course of action and appears to be gaining in strength. These conditions may even persuade the most closed decision units—the insensitive predominant leaders and the single group whose members are highly loyal—to reconsider and change their actions. More probably, though, in the case of the closed decision units, such conditions will lead to a change in the decision unit given authority for this particular problem. The Iran-contra affair is a case in point (Hermann and Hermann, 1990). Reagan's National Security Council staff believed they had been given the authority to get the hostages being held in Lebanon released; the group had "a strong consensus . . . that the President's priority in the Iran initiative was the release of the U.S. hostages" (Tower, Muskie, and Scowcroft, 1987:79–80). They formed a cohesive unit and worked to keep people who disagreed with what they were doing, such as the secretaries of state and defense (Shulz and Weinberger), out of the process. For all intents and purposes, the group, believing they knew what the president wanted and had been given the green light to make the decisions necessary to accomplish the goal, led a covert operation that functioned largely outside the U.S. government, that is, until the story broke and members of the group were held accountable for their actions. At that point the decision unit was forcibly changed.

## Conclusions

We have merely begun the process of making the decision units framework more dynamic. More theoretical and empirical work is needed to elaborate the different ways in which the open and closed decision units can shape or be shaped by what has happened previously in dealing with a particular problem. The expectations decision units have when engaging in an action, the likelihood that an action leaves some members dissatisfied, and the accumulated experiences of the decision unit with the current issue all appear to influence whether or not feedback is monitored, decisions are reconsidered, and the next decision in the sequence involves some change. By setting the policymaking process into motion, we have started to embed the decision units framework into the larger domestic and international environments and moved from portraying a single occurrence to being able to examine a sequence of moves as well as how each impacts the next and emerges from the previous decision. In so doing, it has become clear that the open decision units are more affected by the context in which they find themselves and are interested in both taking advantage of and fitting into what they perceive are relevant constraints. It has also become clear that while often willing to make innovative and risky decisions, the closed decision units are less likely to heed feedback from their environment to move away from the status quo once they have acted. The nature of the decision unit appears to be important in determining how a sequence of actions is going to play out.

All the discussion in this special issue and most of the case studies reviewed in this article have focused on the foreign policy decisions of the governments of states. In the last decade, there has been an explosion of other kinds of actors in the international arena. In fact, some might argue that transnational politics has replaced international politics or, at the least, the two exist side by side—sometimes in conflict with one another (the demonstrations at World Bank or WTO meetings) and at other times living in peaceful coexistence. Among such actors are nongovernmental organizations, international institutions and organizations, regional organizations, multinational corporations, and transnational advocacy networks. Moreover, many important problems in today's world cross borders and are not located within a particular state. Does the decision units framework have applicability to these new actors on the world stage and these new types of problems? Or, more accurately put, can the decision units framework be elaborated to become applicable to such actors and problems?

A number of graduate students whose case studies are not included in the present study because they chose to apply the decision units framework to nonstate actors found it relatively easy to use the framework with other than a state government. They explored such cases as the European Union Council of Ministers decisions regarding imposing sanctions on Austria in response to the election of Jorge Haider, leader of the far right Freedom Party, and the granting of candidacy to Turkey with certain stipulations regarding Cyprus; the Palestinian Liberation Organization decisions regarding the Oslo Accords; decisions to expand the membership of the North Atlantic Treaty Organization; Chechen reactions to the invasion of their territory by Russian forces; and the response of the North American Free Trade Area (NAFTA) to the Mexican peso crisis. Although they found all three types of decision units in the cases at various points in time, the students commented on what they viewed as increasing reliance by such nonstate entities on coalitions and the need to achieve consensus before actions were taken. They also noted the blurring of the lines between what is considered "domestic" and "foreign" and argued for the applicability of the decision units framework to policymaking more generally. Their studies and recommendations are intriguing and invite the development of a set of cases that focus on a wider range of nonstate actors—an important next step for those of us interested in exploring how people and processes are involved in influencing world affairs.

The intent of this special issue was to present a way of understanding how foreign policy decisions are made that emphasizes whose positions count, how such actors are organized, the processes that are likely to result from this organization, and the effects that these processes can have on the decision that results. We have drawn on several decades of research from across the social sciences to propose a contingency approach to studying foreign policymaking that delineates a range of different types of decision units based on their sensitivity to contextual information, where their primary identity lies, and the availability of decision rules. An initial exploration of the accuracy of the framework for explaining historical decisions indicates a rather high degree of overlap between the two. To date the development of the framework has been an iterative process as those involved have moved between theory-building and the examination of historical cases. It is time now for others to join us in applying the framework and helping with its elaboration.

# REFERENCES

Axelrod, Robert (1984) *The Evolution of Cooperation*. New York: Basic Books.

Brecher, Michael (1972) *The Foreign Policy System of Israel*. Oxford: Oxford University Press.

Brockner, J. (1992) The Escalation of Commitment to a Failing Course of Action. *Academy of Management Review* **17**:39–61.

Chen, Shang-Chih, and Jongpil Chung (2000) China's Foreign Policy: 1995–2000. Manuscript, Maxwell School, Syracuse University.

Clark, Russell D., III (1988) On Predicting Minority Influence. *European Journal of Social Psychology* **18**:515–526.

Colvin, I. (1971) *The Chamberlain Cabinet*. London: Victor Gollancz.

Cuhadar, C. Esra (2000) An Application of the Decision Units Framework: The Case of Granting European Union Candidacy to Turkey. Manuscript, Maxwell School, Syracuse University.

De Swaan, Abram (1973) *Coalition Theories and Cabinet Formations*. Amsterdam: Elsevier.

Eagly, Alice H., Wendy Wood, and Shelly Chaiken (1978) Causal Inferences About Communicators and Their Effect on Opinion Change. *Journal of Personality and Social Psychology* **36**:424–435.

Gilboy, George, and Eric Heginbotham (2001) China's Coming Transformation. *Foreign Affairs* **80**:26–39.

Hagan, Joe D. (1993) *Political Opposition and Foreign Policy in Comparative Perspective*. Boulder, CO: Lynne Rienner.

Halperin, Morton H. (1972) The Decision to Deploy the ABM: Bureaucratic and Domestic Politics in the Johnson Administration. *World Politics* **25**:62–95.

Halperin, Morton H. (1974) *Bureaucratic Politics and Foreign Policy*. Washington, DC: Brookings Institution.

Hermann, Charles F., and Robert S. Billings (1993) Sequential Decision Making in Foreign Policy Groups. Paper presented at the annual meeting of the American Political Science Association, Washington, DC, September 2–5.

HERMANN, CHARLES F., AND ROBERT S. BILLINGS (1995) Sequential Group Problem Solving in Foreign Policy: Change or Status Quo? Paper presented at the European Consortium for Political Research, Bordeaux, France, April 27–May 2.

HERMANN, MARGARET G., AND CHARLES F. HERMANN (1989) Who Makes Foreign Policy Decisions and How: An Empirical Inquiry. *International Studies Quarterly* **33**:361–387.

HERMANN, MARGARET G., AND CHARLES F. HERMANN (1990) "Hostage Taking, the Presidency, and Stress." In *Origins of Terrorism: Psychologies, Ideologies, Theologies, States of Mind*, edited by Walter Reich. Cambridge: Cambridge University Press.

HOYT, PAUL D., AND JEAN A. GARRISON (1997) "Political Manipulation Within the Small Group: Foreign Policy Advisers in the Carter Administration." In *Beyond Groupthink: Political Group Dynamics and Foreign Policymaking*, edited by Paul 't Hart, Eric K. Stern, and Bengt Sundelius. Ann Arbor: University of Michigan Press.

KAARBO, JULIET (1998) Power Politics in Foreign Policy: The Influence of Bureaucratic Minorities. *European Journal of International Relations* **4**:67–97.

KAARBO, JULIET, AND RYAN K. BEASLEY (1998) "A Political Perspective on Minority Influence and Strategic Group Composition." In *Research on Groups and Teams*, vol. 1, edited by Margaret A. Neale, Elizabeth Mannix, and Deborah H. Gruenfeld. Greenwich, CT: JAI Press.

KAARBO, JULIET, AND DEBORAH H. GRUENFELD (1998) The Social Psychology of Inter- and Intragroup Conflict in Governmental Politics. *Mershon International Studies Review* **42**:226–233.

LINDBLOM, CHARLES E. (1959) The Science of Muddling Through. *Public Administration Review* **19**:79–88.

MIDDLEMAS, K. (1972) *The Diplomacy of Illusion*. London: Weidenfeld and Nicolson.

MOSCOVICI, SERGE (1985) "Innovation and Minority Influence." In *Perspectives on Minority Influence*, edited by Serge Moscovici, Gabriel Mugny, and E. Van Avermaet. Cambridge: Cambridge University Press.

MUGNY, GABRIEL (1975) Negotiations, Image of the Other, and the Process of Minority Influence. *European Journal of Social Psychology* **5**:209–228.

MUGNY, GABRIEL, AND JUAN A. PEREZ (1991) *The Social Psychology of Minority Influence*. Cambridge: Cambridge University Press.

OZKECECI, BINNUR (2000) Turkey, Greece, and the Cyprus Crisis of 1974. Manuscript, Maxwell School, Syracuse University.

SALANCIK, G. R., AND J. R. MEINDL (1984) Corporate Attributions as Strategic Illusions of Managerial Control. *Administrative Science Quarterly* **29**:238–254.

SNYDER, RICHARD C., H. W. BRUCK, AND BURT SAPIN (1962) *Foreign Policy Decision Making*. New York: Free Press.

STAW, B. M., AND J. ROSS (1987) Behavior in Escalation Situations: Antecedents, Prototypes, and Solutions. *Research in Organizational Behavior* **9**:39–78.

STEINBRUNER, JOHN D. (1974) *The Cybernetic Theory of Decision*. Princeton, NJ: Princeton University Press.

STERN, ERIC K. (1999) *Crisis Decisionmaking: A Cognitive-Institutional Approach*. Stockholm: Department of Political Science, Stockholm University.

STEVENSON, M. K., J. R. BUSEMEYER, AND JAMES C. NAYLOR (1990) "Judgment and Decision-Making Theory." In *Handbook of Industrial and Organizational Psychology*, edited by M. D. Dunnette and L. M. Hough. Palo Alto, CA: Consulting Psychologists Press.

STEWART, PHILIP D., MARGARET G. HERMANN, AND CHARLES F. HERMANN (1989) Modeling the 1973 Soviet Decision to Support Egypt. *American Political Science Review* **83**:35–59.

SUNDELIUS, BENGT, ERIC STERN, AND FREDRIK BYNANDER (1997) *Crisis Management the Swedish Way—In Theory and Practice*. Stockholm: Swedish Agency for Civil Emergency Planning.

SWORDS, DIANE R. (2000) U.S. and Iraqi Decisions in February 1998 Weapons Inspection Crisis. Manuscript, Maxwell School, Syracuse University.

TOWER, JOHN, EDMUND MUSKIE, AND BRENT SCOWCROFT (1987) *The Tower Commission Report*. New York: Bantam Books.

UHLER, BETH D. (1993) The Escalation of Commitment in Political Decision-Making Groups. Ph.D. dissertation, University of Pittsburgh.

VALENTA, JIRI (1979) The Bureaucratic Politics Paradigm and the Soviet Invasion of Czechoslovakia. *Political Science Quarterly* **94**:55–76.

VERTZBERGER, YAACOV Y. (1990) *The World in Their Minds*. Stanford, CA: Stanford University Press.

WALKER, STEPHEN G. (1990) British Decision-Making and the Munich Crisis. Paper presented at the annual meeting of the International Studies Association, Washington, DC, March.

WALKER, STEPHEN G., AND GEORGE L. WATSON (1989) Groupthink and Integrative Complexity in British Foreign Policymaking: The Munich Case. *Cooperation and Conflict* **24**:199–212.

WONG, P. T., AND B. WEINER (1981) When People Ask "Why" Questions and the Heuristics of Attributional Search. *Journal of Personality and Social Psychology* **40**:650–663.

WOOD, W., S. LUNDGREN, J. A. OULETTE, S. BUSCEME, AND T. BLACKSTONE (1994) Processes of Minority Influence: Influence Effectiveness and Source Perceptions. *Psychological Bulletin* **115**:323–345.

# APPENDIX: CASE STUDIES

| Country | Date | Case |
|---|---|---|
| Argentina | 1978 | Inspection visit of Inter-American Commission on Human Rights regarding violations |
| | 1982 | Decision to invade the Falkland Islands |
| | 1984 | Ratification of Beagle Channel Treaty |
| Brazil | 1991 | Pressures to preserve the Amazon jungle |
| Britain | 1938 | Munich crisis |
| | 1956 | Decisions regarding the Suez crisis |
| | 1982 | Response to Falkland Islands invasion |
| Canada | 1969 | Reduction of commitment to NATO |
| China | 1962 | Border dispute with India |
| | 1990 | Negotiations on purchase of advanced Soviet aircraft |
| | 1995 | Taiwanese president's visit to the United States |
| | 1999 | Negotiations with the U.S. over entry into the World Trade Organization |
| Colombia | 1997 | Negotiations over demilitarized zone in FARC territory |
| | 1999 | Peace talks with FARC and ELN |
| Egypt | 1962 | Intervention into Yemen |
| | 1973 | Decision to go to war with Israel |
| France | 1959 | Recognition of Algerian independence |
| | 1991 | Breakup of Yugoslavia |
| Germany | 1970 | Ostpolitik negotiations |
| Greece | 1974 | Turkish intervention into Cyprus |
| India | 1962 | Forward Policy and Operation Leghorn |
| | 1967 | Refusal to sign Nonproliferation Treaty |
| | 1971 | Involvement in Pakistani Civil War |
| Indonesia | 1998 | Reactions to demands of IMF during Asian financial crisis |
| Iran | 1979 | Taking of American hostages |
| | 1995 | U.S. blocks Conoco oil deal |
| | 1999 | Visits to Western Europe seeking trade and investment |
| Iraq | 1996 | Breach of cease-fire between Kurdish groups |
| | 1997 | "Cat and mouse" game with United Nations weapons inspection team |

| Country | Date | Case |
|---|---|---|
| Israel | 1967 | Series of threats by Egypt and other Arab states |
| | 1976 | Raid on Entebbe |
| | 1982 | Invasion into southern Lebanon |
| | 1991 | Reaction to Iraqi bombing |
| Japan | 1969 | Reactions to the "textile wrangle" with the United States |
| | 1971 | Response to "Nixon shocks" and pressure to devalue the yen |
| | 1972 | Tyumen Development Project |
| | 1992 | Negotiations surrounding the Peacekeeping Operations Bill |
| Mexico | 1962 | Organization of American States' vote on Cuba |
| Netherlands | 1980s | Acceptance of NATO cruise missiles |
| New Zealand | 1985 | Negotiations regarding visiting rights for U.S. ship |
| Nigeria | 1975 | Recognition of the MPLA |
| South Africa | 1989 | Decisions regarding "shadow membership" in nuclear club and nuclear rollback |
| South Korea | 1998 | "Sunshine Policy" toward North Korea |
| Soviet Union/Russia | 1956 | Invasion of Hungary |
| | 1961 | Construction of Berlin Wall |
| | 1968 | Invasion of Czechoslovakia |
| | 1973 | Negotiations regarding providing offensive weapons to Egypt |
| Sweden | 1981 | Soviet submarine trapped on the rocks within Swedish territory |
| Syria | 1976 | Intervention into Lebanese civil war |
| | 1987 | Tensions with Hezbollah in Lebanon |
| Taiwan | 1999 | Decision to declare "two-states theme" |
| Tanzania | 1978 | Invasion by Ugandan forces and annexation of Ugandan territory |
| Turkey | 1974 | Ouster of Cyprus President Makarios |
| | 1999 | Response to offer of EU candidacy with conditions |
| Ukraine | 1992 | Decision to sign the Lisbon Protocols to the START I Treaty |
| United States | 1961 | Reactions to tensions in Berlin |
| | 1965 | Escalation of U.S. involvement in Vietnam |
| | 1980 | Rescue mission of hostages in Teheran |
| | 1982 | Decisions regarding the U.N. Convention on the Law of the Sea |
| | 1986 | Dispute with Japan over semiconductor trade |
| | 1986 | Bombing of Libya |
| | 1990 | Iraqi invasion of Kuwait |
| | 1992 | Development of the Lisbon Protocols to the START I Treaty |
| Yugoslavia | 1995 | Negotiations leading to the Dayton Accords |
| | 1999 | Negotiations over Kosovo at Rambouillet |

**THE INTERNATIONAL STUDIES REVIEW** (ISSN 1521-9488) is published three times a year, in spring, summer, and fall, by Blackwell Publishers, Inc., with offices at 350 Main Street, Malden, MA 02148, USA, and 108 Cowley Road, Oxford OX4 1JF, UK. Call USA 1-800-835-6770 or (781) 388-8200, fax: (781) 388-8232, e-mail: subscrip@blackwellpub.com. Visit us online at www.blackwellpub.com. Subscription also includes four issues of *International Studies Quarterly* and four issues of *International Studies Perspectives*.

**INFORMATION FOR SUBSCRIBERS** New orders, renewals, sample copy requests, claims, change of address information and all other correspondence should be sent to the Subscriber Services Coordinator at the publisher's Malden office (see address above).

**SUBSCRIPTION RATES FOR VOLUME 3, 2001**

|  | North America | Rest of the World |
|---|---|---|
| Subscription: | $480 | £319 |
| Single Issues: | $52 | £35 |

Individual members of the International Studies Association receive a subscription to the *Quarterly* and the *Review* as one of the many benefits of membership. Contact the ISA at Social Science 324, University of Arizona, Tucson, AZ 85721, USA. Tel.: (520) 621-1208.

Checks and money orders should be made payable to Blackwell Publishers. Canadian residents please add 7% GST. Checks in US Dollars must be drawn on a US bank. Checks in Sterling must be drawn on a UK bank. All orders must be paid by check, money order, or credit card. Only personal orders may be paid by personal check.

**BACK ISSUES** Single issues from the current volume are available from the publisher's Malden office at the current single issue rate.

**MICROFORM** The journal is available on Microfilm. For microfilm service address inquiries to University Microfilms Library Services, 300 North Zeeb Road, Ann Arbor, MI 48106-1346, USA.

**MAILING** The journal is mailed second class in North America and by IMEX to the rest of the world.

**POSTMASTER** Third-class postage pending at Boston, MA and additional offices. Send all address corrections to Blackwell Publishers, Journals Subscriptions Department, 350 Main Street, Malden, MA 02148.

**ADVERTISING** For information and rates, please contact Publishers Communication Group. Phone (617) 395-4055, Fax (617) 354-6875, or e-mail sabine.mourlon@pegplus.com.

For all other permissions inquiries, including requests to republish material in another work, please contact the Journals Permissions Manager at the publisher's Oxford office, 108 Cowley Road, Oxford, OX4 1JF, U.K., te: +44 1865 791100, fax: +44 1865 791347.

**INDEXING/ABSTRACTING** The contents of this journal are indexed or abstracted in the following: *ABC POL SCI, Academic Abstracts, America: History & Life, American Bibliography of Slavic & Eastern European Studies, British Humanities Index, Current Contents/Social & Behavioral Sciences, Current Index to Journals in Education, Geographical Abstracts, Historical Abstracts, Human Resources Abstracts, Human Rights Internet Reporter, Index of Islamic Literature, International Bibliography of Sociology, International Development Abstracts, International Political Science Abstracts, Journal of Economic Literature, Middle East: Abstracts & Index, Newspaper & Periodical Abstracts, Public Affairs Information Service, Peace Research Abstracts Journal, Poverty & Human Resources Abstracts, Predicasts Forecasts, Research Alert, Risk Abstracts, Sage Public Administration Abstracts, Social Sciences Citation Index, Social Science Index, Studies on Women Abstracts, United States Political Science Documents,* and *Western Historical Quarterly.*

© 2001 International Studies Association

This special issue of the ISR, while published under joint sponsorship of the International Studies Association and Blackwell Publishers, does not necessarily follow the same editorial guidelines as the regular issues of the ISR. Scholars who wish to contribute to the regular issues should contact the ISR editorial office: *International Studies Review*, The Watson Institute for International Studies, Brown University, Box 1970, 2 Stimson Ave., Providence, RI 02912-1970 USA. Phone: 401-863-2809, Fax: 401-863-1270, E-mail: ISR@brown/edu.

## INTERNATIONAL STUDIES REVIEW

*International Studies Review* enhances the understanding of the theory and practice of international relations in global and regional settings by publishing the following:

**Reflection and Reappraisal.** Essays of 8,000–10,000 words that analyze current scholarship and recent advances in core subfields. Especially welcome are contributions that assess the state of knowledge beyond North American perspectives or incorporate the insights of history, sociology, anthropology, geography, and economics as well as political science.

**Review Essays.** Essays of 4,000–5,000 words that place three or more recently published volumes in a broad conceptual context and suggest future research agenda.

**Reviews.** Essays of 1,000–2,000 words that clarify the value of one or two books to contemporary debates or highlight non-English language works.

**Response and Reaction.** Authors may submit responses of no more than 1,000–2,000 words to material previously published in *ISR*.

We welcome submissions and urge authors to initiate the process by sending a 2–5 page proposal that indicates the type of essay being prepared together with the literature being reviewed. When essays are commissioned, the author will receive style sheets from the editor. Completed essays will be refereed.

Inquiries should be sent to: Frederick Fullerton, Managing Editor, *International Studies Review*, The Watson Institute for International Studies, Brown University, Box 1970, 2 Stimson Avenue, Providence, RI 02912-1970 USA. Phone: 401-863-2809, Fax: 401-863-1270, E-mail: Frederick_Fullerton@brown.edu.

## New from **Columbia**

### The Ties That Divide
Ethnic Politics, Foreign Policy, and International Conflict
**Stephen M. Saideman**

Applying his insights to the Congo Crisis, the Nigerian Civil War, Yugoslavia's civil wars, and finally the recent events in Kosovo, Saideman argues that domestic political competition compels countries to support the side of an ethnic conflict with which its constituents share ethnicities.
348 pages • 37 tables, 8 graphs • $19.50 paper

### Violent Peace
Militarized Interstate Bargaining in Latin America
**David R. Mares**

"A brilliant synthesis of theory and historical case studies that characterized the 'violent peace' of Latin America, this book advances the study of inter-American security to a new level of sophistication. It should be required reading among senior diplomatic and military officials and university courses on inter-American relations."—Gabriel Marcella, U.S. Army War College
398 pages • 28 tables, 7 figures • $19.50 paper

### Problematic Sovereignty
Contested Rules and Political Possibilities
**Edited by Stephen D. Krasner**

"*Problematic Sovereignty* addresses issues of great contemporary importance in world politics in a fresh and provocative way and artfully engages theoretical debates with in-depth, very high-quality case studies."—Daniel Deudney, Johns Hopkins University
502 pages • $22.50 paper

### Globalization and the European Political Economy
**Edited by Steven Weber**

What mix of causal forces from global and national levels accounts for the changes within European government? Is Europe "special" in any sense? Weber and and his contributors explore the impact of globalization on governance structures in the modern European political economy.
378 pages • $19.50 paper

### Gendering World Politics
Issues and Approaches in the Post-Cold War Era
**J. Ann Tickner**

Expanding on the issues she originally explored in her classic work, *Gender in International Relations*, J. Ann Tickner focuses her distinctively feminist approach on new issues of the international relations agenda since the end of the Cold War, such as ethnic conflict and other new security issues, globalization, democratization, and human rights.
262 pages • $17.50 paper

## COLUMBIA UNIVERSITY PRESS
columbia.edu/cu/cup    800-944-8648

# SOCIAL SCIENCE QUARTERLY

PUBLISHED ON BEHALF OF THE SOUTHWESTERN SOCIAL SCIENCE ASSOCIATION

Edited by ROBERT L. LINEBERRY

"Good coverage of contemporary social questions from a research standpoint." *Magazines for Libraries*

Nationally recognized as one of the top journals in the field, *Social Science Quarterly* (SSQ) publishes current research on a broad range of topics including political science, economics, history, geography, and women's studies.

ISSN: 0038-4941. VOLUME 82 (2001) CONTAINS 4 ISSUES.
WWW.BLACKWELLPUBLISHERS.CO.UK/JOURNALS/SSQ

**Select** ✓ BLACKWELL PUBLISHERS' EMAIL UPDATES

WE ARE PLEASED TO ANNOUNCE THE LAUNCH OF OUR NEW EMAIL ALERTING SERVICE. YOU CAN NOW RECEIVE THE TABLES OF CONTENTS OF **SOCIAL SCIENCE QUARTERLY** EMAILED DIRECTLY TO YOUR DESKTOP. UNIQUELY FLEXIBLE, SELECT ALLOWS YOU TO CHOOSE EXACTLY THE INFORMATION YOU NEED. FOR FREE UPDATES ON BLACKWELL PUBLISHERS' TITLES SIMPLY VISIT:
HTTP://SELECT.BLACKWELLPUBLISHERS.CO.UK

✓ **SELECT** EXACTLY WHAT YOU WANT TO RECEIVE
✓ **SELECT** CONTENTS TABLES FROM THE JOURNALS OF YOUR CHOICE
✓ **SELECT** NEWS OF BOOKS AND JOURNALS BY SUBJECT AREA
✓ **SELECT** WHEN YOUR MESSAGES ARRIVE, RIGHT DOWN TO THE DAY OF THE WEEK

108 COWLEY ROAD, OXFORD OX4 1JF, UK
350 MAIN STREET, MALDEN, MA 02148, USA
JNLINFO@BLACKWELLPUBLISHERS.CO.UK

VISIT OUR WEBSITE FOR CONTENTS LISTINGS, ABSTRACTS, SAMPLES, AND TO SUBSCRIBE

W W W . B L A C K W E L L P U B . C O M

# World Press Review takes these...

## and gives you this...

There's no need for you or your students to read through a myriad of newspapers and magazines. We've already done that for you! **WORLD PRESS REVIEW** serves as a window into other societies by sifting through and translating hundreds of articles from newspapers and magazines from all over the world, looking for items that illuminate, instruct, and entertain. **WORLD PRESS REVIEW** is perfect for teaching economics, political science, government, journalism, or international relations, and is one of the most valuable reference sources available.

Our order schedule is flexible. As an added benefit, you'll be able to customize your subscription order to coincide with your school semester. And, when 10 or more student subscriptions are ordered, the educator receives a free desk copy. For more information log on to our website: **www.worldpress.org**

Order **WORLD PRESS REVIEW** today, or mail us this coupon for a free sample copy.

**YES, I WANT MY FREE COPY OF WORLD PRESS REVIEW!**

............................................................................

❏ Mr./ ❏ Ms.: _____

Address: _____

City: _____ State: _____ Zip: _____

Email (optional): _____

❏ Student    ❏ Educator   (check if appropriate)

**Mail today to:**
**WORLD PRESS** Review
700 Broadway, New York, NY 10003
(212) 982-8880 X100
www.worldpress.org

(ISRFALL01)

Printed and bound by CPI Group (UK) Ltd, Croydon, CR0 4YY
09/06/2025

14685997-0003